Lab Manual to Accompany

INTRODUCTION TO FORESTRY SCIENCE

Third Edition

Renee Peugh
L. DeVere Burton

DELMAR
CENGAGE Learning

Australia • Brazil • Japan • Korea • Mexico • Singapore • Spain • United Kingdom • United States

DELMAR
CENGAGE Learning

Lab Manual to Accompany Introduction to Forestry Science, Third Edition
L. DeVere Burton, Renee Peugh

Vice President, Editorial: Dave Garza
Director of Learning Solutions: Matthew Kane
Senior Acquisitions Editor: Sherry Dickinson
Managing Editor: Marah Bellegarde
Senior Product Manager: Christina Gifford
Editorial Assistant: Scott Royael
Vice President, Marketing: Jennifer Baker
Marketing Director: Debbie Yarnell
Marketing Manager: Erin Brennan
Marketing Coordinator: Erin DeAngelo
Senior Production Director: Wendy Troeger
Production Manager: Mark Bernard
Senior Content Project Manager: Katie McGuire
Senior Art Director: David Arsenault

Library of Congress Control Number: 2011941346

ISBN-13: 978-1-111-30841-4
ISBN-10: 1-111-30841-1

Delmar
5 Maxwell Drive
Clifton Park, NY 12065-2919
USA

Cengage Learning is a leading provider of customized learning solutions with office locations around the globe, including Singapore, the United Kingdom, Australia, Mexico, Brazil, and Japan. Locate your local office at: **international.cengage.com/region**

Cengage Learning products are represented in Canada by Nelson Education, Ltd.

To learn more about Delmar, visit **www.cengage.com/delmar**
Purchase any of our products at your local college store or at our preferred online store **www.cengagebrain.com**

Printed in the United States of America
1 2 3 4 5 18 17 16 15 14

CONTENTS

PREFACE

The overall purpose of the Lab Manual to accompany *Introduction to Forestry Science,* Third Edition, is to provide students with practical, "hands-on" activities that will reinforce concepts taught in the text. Research has shown that students learn best when instructors use a variety of approaches. The acquisition of cognitive skills in the classroom should be followed by laboratory activities that allow students to apply the concepts.

The activities provided in this Lab Manual help teach students the value of our forests as a natural resource and also the scientific principles behind the growth, development, and management of the forests. Few topics are as important as the stewardship and management of the forest, not only as a natural resource, but also as an agricultural crop. By presenting these concepts, instructors have an opportunity to teach science principles in an applied setting. Research clearly documents that contextual learning is one of the most effective means to ensure student achievement. The activities in this manual help teach the contextual and applied facets of science.

The Lab Manual that accompanies the text is designed to help the instructor prepare lab activities for each of the 20 chapters in the text. Each chapter begins with the student objectives, followed by activities designed to achieve the objectives. When appropriate, exercises follow the scientific method and require students to create a hypothesis, record observations, analyze data, and draw conclusions.

Chapter 1

INTRODUCTION TO FORESTRY—*PLEASED TO MAKE YOUR ACQUAINTANCE*

OBJECTIVES

After completing these lab exercises, students should be able to:

- discuss the wide range of products that come from forests
- explain the difference between a primary product and a by-product
- recognize everyday products that come from the forest
- explain how the forest helps the environment
- discuss the effects of trees on temperature
- explain how trees cool the temperature
- determine wildlife use of an area

EXERCISE 1 HEY MOM, LOOK WHAT I MADE—FORESTRY PRODUCTS

Background

Forests are an important asset to people all over the world. We benefit from the biological processes of the plants and trees, as well as the products produced from them. Without forests we would have a much lower quality of life because we would have fewer consumer products, recreational opportunities, and employment options, not to mention a less healthy environment. All around our homes are products that come directly or indirectly from the forest. Many of these we use every day and take for granted.

Procedures

Using the table that follows, make an inventory of all the forestry products used in and around your home. In the first column, list the object or product; and in the second, list the source of the product. Then, in the third column, indicate whether the product is a primary product or a by-product. A **primary product** is one that is grown and/or manufactured from trees. Lumber is a primary product, as are other items made from raw lumber. A **by-product** is one produced during the process of growing or manufacturing a primary product. For example, sawdust is a by-product of lumber. The source is the part of a tree(s) from which a consumer product is made.

Observations

Collect Data

Inventory of Forest Products Used In and Around the Home

Product	Source	Primary or By-Product
Kitchen table	Lumber	Primary product
Mulch	Pine bark	By-product

Product	Source	Primary or By-Product

Conclusions

Prepare and present an oral report to the class on the products you found in and around your home. Prepare an estimate of the value of these products. Be sure to emphasize any unusual or especially interesting products you discovered. Hint: The toothpaste you use may have ingredients that came from forestry products.

EXERCISE 2 THERE IS METHOD TO THE MADNESS—THE SCIENTIFIC METHOD

Background

The **scientific method** is the process scientists use as they explore the world around them. There are six basic steps in this process. In this manual, some labs will require you to use all steps, but many labs will only focus on one or only a few. It is important for forestry students to understand the scientific method. As you explore and use these methods of discovery, you will gain a greater appreciation and understanding of the forests.

Materials Needed

Students will determine the materials they need for this exercise.

Procedures

Review the steps of the scientific method below. In small groups, you will use these steps to create and carry out an experiment of your own.

Step 1 *Observation*: Using any or all of your senses, observe something that peaks your curiosity. Keep you focus on forestry. Then ask the question, "What do I want to know?"
Step 2 *Research:* Find out what others have already discovered. Working scientists primarily use peer-reviewed journals as their main sources of information. You will use the Internet and periodicals for your research. Just be sure your information comes from credible resources.
Step 3 *Hypothesis:* Form a hypothesis, an educated guess, about what the answer to your question is.
Step 4 *Experiment:* Create and carry out an experiment that will test your hypothesis.
Step 5 *Collect data:* Data will include and any information you learned from your experiment. It may include measurements, drawings, or any information you observed.
Step 6 *Conclusions:* Review your data, and summarize the results. You may need to make calculations and then form a judgment about your hypothesis. Do your findings support or reject your hypothesis? This is also the time when you would write a report and create figures and diagrams to illustrate your conclusions.

Step 1 ***Observation.*** In this lab, the problem has already been identified. Using the remaining steps of the scientific method, complete the following exercise:

Problem: You notice that dark or shiny items seem hotter in the sun than light nonreflective items. You wonder how an item's color and reflective ability affect its temperature and its ability to absorb or to reflect light.

Step 2 ***Research.*** Find out what others have learned. In the space below, list any reference you used for your research and a brief summary of what was learned.

____

____

____

____

____

____

Step 3 ***Hypothesis.*** Form a hypothesis and write it in the space provided.

____

____

Step 4 ***Experiment.*** Design an experiment that will test your hypothesis. Include a detailed explanation of your experiment. Be sure to include a timeline, estimated cost, equipment, and any other important information about your experiment here.

____

____

____

Step 5 ***Collect Data.*** Record all information you collected in the space below or design your own data format. Be sure to include any pictures or drawings if applicable.

Step 6 ***Conclusions.*** Write a brief report summarizing your results; accept or reject your hypothesis.

EXERCISE 3 IS IT HOT OUT HERE?—TREES AND TEMPERATURE

Background

Forests are an essential part of our environment and our ecological system. Trees help cool the temperature; provide food, shade, and homes for wildlife; and put oxygen into the air. They also help lessen the effects of air pollution.

Materials Needed

- 2 large outdoor thermometers
- Large clear plastic bags and ties

Procedures

This lab exercise is designed to be used in the fall or spring when the weather is warm and the trees have leaves on them. Hang a thermometer in the shade under a large deciduous tree. Hang another in the sun away from the tree. For five days, each day at the same time (preferably around noon), record the temperatures registered on the thermometers under the tree and in the sun, using the table that follows. At the end of five days, average the temperatures recorded in each setting and compare them.

Next select a limb on the tree and, in the middle of the day, place a clear plastic bag around the leaves on the end of the limb (see Figure 1–1). In about two hours, carefully remove the plastic bag and measure the amount of water in the bag. What effect does this have on the temperature? Try to count the number of limbs on the tree that are the same size as the part covered by the bag. From this estimate, determine approximately how much water the tree gives off in an hour.

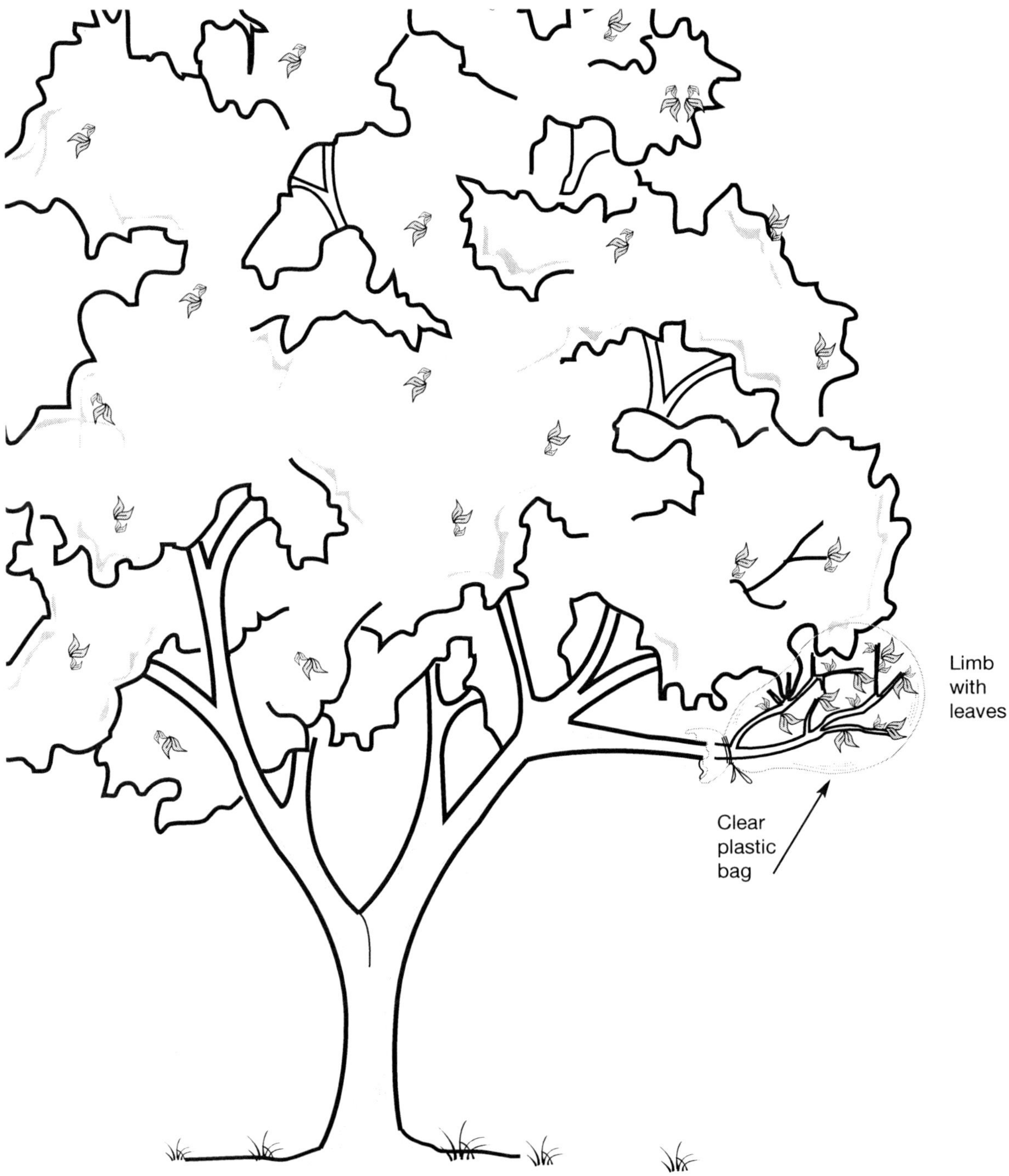

Figure 1–1

Hypothesis

Write a brief statement predicting what you think you will happen.

____

____

Observations

Collect Data

	Temperature under Tree	Temperature in Sun
Day 1		
Day 2		
Day 3		
Day 4		
Day 5		
Average		

	Temperature under Tree	Temperature in Shade of Building
Day 1		
Day 2		
Day 3		
Day 4		
Day 5		
Average		

Conclusions and Further Action

The average temperature under the tree should be cooler than the temperature in the sun. You should now begin to try to discover why this is true. An obvious answer is that the shade under the leaves causes the difference. Try this experiment again, but instead of hanging the thermometer in the sun, hang it in the shade of a board or building and record the differences in temperature on the table. You should find that there are differences in the temperature between the thermometers in the shade of the tree and in the shade of the board or building.

Write a statement summarizing what you found. Then accept or reject your hypothesis.

____

____

____

____

EXERCISE 4 I JUST NEED A LITTLE FRESH AIR—TREES AND AIR QUALITY

Background

Trees can help reduce air pollution. Particles in the air are often trapped on the leaves of trees and adhere there until the rain washes off the pollution. Through this action, forests can act as a filter against air pollution.

Materials Needed

- 10 strips of cardboard measuring 4 × 12 inches
- A mathematical compass
- Scissors heavy enough to cut cardboard
- 1 roll of 2-inch clear packing tape
- Several lengths of string (2-feet long)

Procedures

This exercise should be done in the fall or spring when the weather is warm and the trees have leaves on them. Using the compass, draw four evenly spaced circles 1½ inches in diameter on the strips of cardboard. Next, cut out the circles so that you have four, 1½-inch holes in the cardboard. On the back of the cardboard, place a strip of the tape so that all the holes are covered. Avoid touching the sticky side (see Figure 1–2). Find several places around the school campus where you can hang the cardboard with the tape. Suspend the strips about 4 feet off the ground, using the string. Place half the strips in openings away from buildings and trees, and place the other half under trees and shrubs among the leaves (see Figure 1–3a and 1–3b).

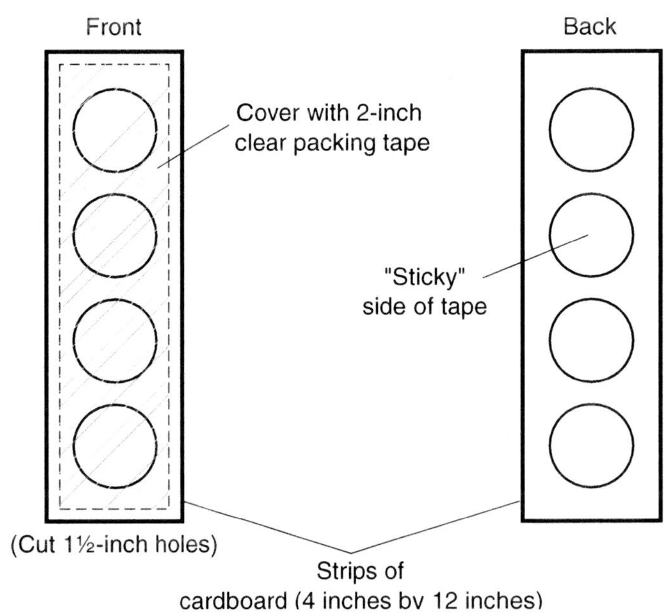

Figure 1–2

Figure 1–3a

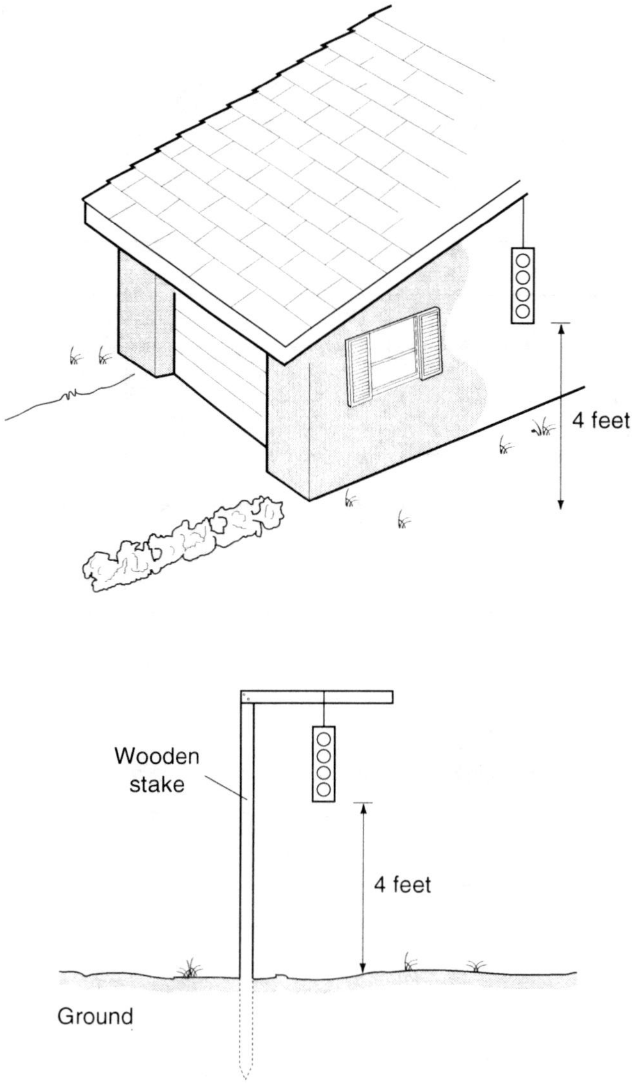

Figure 1–3b

Leave the strips suspended for 24 hours and then collect them. Be sure to mark on the cardboard where the strips were located (under trees or in the open). Using a magnifying glass or microscope, examine the sticky areas for any materials adhering to the surface. Using the table that follows, record the number of specks that you observe on each strip.

Next, total the number of particles that you found on the strips placed in open spaces and then the number on the strips placed under trees. Compare the figures and draw a conclusion. You might also want to attempt to determine the source of the air pollution.

Hypothesis

Write a brief statement predicting what you think you will find.

Observations

Collect Data

	Strip Number	Location	Number of Particles	Type of Particle
1				
2				
3				
4				
5				
6				
7				
8				
9				
10				

Conclusions

Review your data for a conclusion. Write a summary statement describing what you found.

Chapter 2

NORTH AMERICAN FOREST REGIONS—*CAN'T SEE THE FOREST FOR THE TREES*

OBJECTIVES

After completing this lab exercise, students should be able to:

- cite pertinent facts concerning our forests
- find information on our forests
- distinguish fact from fiction regarding our forests

EXERCISE 1 THAT'S A LIE—DISTINGUISHING FACT FROM FICTION

Background

Most people would probably say that they have an affection for the forest. An integral part of our history, the forest represents a large portion of our economy. Many groups work toward trying to preserve forest trees as an essential part of our environment. Foresters and loggers see trees not only as a part of our environment, but also as a means to make a living. Groups often disagree over management practices or even over whether we should have management practices. Sometimes people can perceive information wrongly, and sometimes information is wrong. As always, a thorough search for facts can be beneficial. In this exercise, you will learn to distinguish fact from fiction.

Materials Needed

- Access to a computer and the Internet

Procedures

Several statements have been provided regarding an opinion or perception about forests in this country. Below each statement is a place for you to indicate whether the statement is a fact (truth) or fiction (false). Circle the correct answer. You then have space in which to back up your decision with facts gained from an Internet search. Be thorough in your search and try to gain information from several sources. If any of your classmates disagree with you concerning the validity of a statement, perhaps a debate can be held in the classroom. Remember to base your thoughts and opinions on information from reliable sources. As you search the Internet, enter words such as "forests," "forest ecology," "wilderness areas," "virgin timber," "old-growth trees," and so on.

Hypothesis

Write a brief statement predicting what you believe you will discover. Do you think most common beliefs about our forests are fact or fiction?

____

____

Assertion #1

The early U.S. forest was a carpet of trees that extended from coast to coast. *Note:* You may want to do image searches and see for yourself.

FACT or FICTION

Supporting facts: ____

____

____

____

____

Assertion #2

We have only 5 percent of the original old-growth forests left that once covered the Pacific Northwest in the pre-European settlement era.

FACT or FICTION

Supporting facts: ____________________

Assertion #3

In authorizing salvage logging of dead and dying timber, Congress ignored environmental safeguards.

FACT or FICTION

Supporting facts: ____________________

Assertion #4

We are running out of trees.

FACT or FICTION

Supporting facts: ____________________

Assertion #5

We are cutting more trees than we are growing for future generations.

FACT or FICTION

Supporting facts: ____________________

Assertion #6

We are running out of wilderness.

FACT or FICTION

Supporting facts: ____________________

Assertion #7

Clear cutting, the practice of harvesting most trees in a given area, destroys the forest.

FACT or FICTION

Supporting facts: ____

____

____

____

____

Assertion #8

A natural forest supports more ecological diversity than a managed forest.

FACT or FICTION

Supporting facts: ____

____

Assertion #9

Forest management harms fragile wetlands.

FACT or FICTION

Supporting facts: ____

____

____

____

____

Assertion #10

Forest management harms wildlife.

FACT or FICTION

Supporting facts: ____

____

____

____

____

Assertion #11

More paper recycling will help us avoid the use of "virgin" wood from harvested trees.

FACT or FICTION

Supporting facts: ____

____

____

____

____

Assertion #12
Forest product companies ignore concerns about wildlife.

FACT or FICTION

Supporting facts: ______________________________

Assertion #13
The forest industry employs a small number of people.

FACT or FICTION

Supporting facts: ______________________________

Assertion #14
Wildlife populations have decreased since the beginning of the twentieth century because of poor forest management.

FACT or FICTION

Supporting facts: ______________________________

Assertion #15
Only one seedling per American is planted every year, and environmental groups plant most seedlings.

FACT or FICTION

Supporting facts: ______________________________

Assertion #16
Private owners control the main portion of forested land in the United States.

FACT or FICTION

Supporting facts: ____________________

____

____

____

____

Assertion #17

In the United States, more paper is used to publish books than magazines and newspapers.

FACT or FICTION

Supporting facts: ____________________

____

____

____

____

Assertion #18

About one-half of the United States is forested.

FACT or FICTION

Supporting facts: ____________________

____

____

____

____

Assertion #19

Georgia is ranked number 1 in the number of acres of commercial forest land.

FACT or FICTION

Supporting facts: ____________________

____

____

____

____

Assertion #20

The United States preserves old-growth forest land in various ways.

FACT or FICTION

Supporting facts: ____________________

____

____

____

____

Assertion #21

The average single-family home (2,000 square feet) contains the same number of board feet lumber (that is, 2,000 board feet of lumber).

FACT or FICTION

Supporting facts: ____________________

____

____

____

____

Assertion #22

Wood is cheaper in total energy costs than other competitive energy sources.

FACT or FICTION

Supporting facts: ____________________

____

____

____

____

Assertion #23

Paper and paperboard recovery has more than doubled.

FACT or FICTION

Supporting facts: ____________________

____

____

____

____

Assertion #24

The forest industry is composed of paper mills only.

FACT or FICTION

Supporting facts: ____________________

____

____

____

____

Assertion #25

America's forest product companies have donated 500,000 acres of land for public use.

FACT or FICTION

Supporting facts: ____

____

____

____

____

Conclusions

Write a statement summarizing your findings. Were most of the common belief statements facts or fiction?

____

____

____

____

Chapter 3

SAFETY IN THE FOREST—*IT'S NO ACCIDENT*

OBJECTIVES

After completing these lab exercises, students should be able to:

- recognize hazardous areas and working conditions
- determine their relative reaction time
- compare their reaction time to that of machinery
- properly identify areas that should be color-coded
- recognize specific hazards indicated by specific colors

EXERCISE 1 REACTION TIME AND CONSEQUENCES

Background

Many of the jobs associated with the forest industry are dangerous, and accidents can result in serious injury or death. Removing felled trees, trimming them into logs, and removing the logs from the forest all involve the use of powerful machines. Even the processing plants where the trees are made into usable products are hazardous because of all the machinery involved. A thorough understanding of how machines operate and the danger points of such operations is essential in preventing injuries. Most machines are powerful and have parts that move much more rapidly than humans can. People sometimes forget that their reaction time is slow in comparison to that of machinery. The following exercise will help illustrate how slowly we react to machinery.

Materials Needed

- Yardstick
- Stopwatch

Procedures

To test reaction time, have one student place his or her hand flat on the edge of a desk. Have a second student hold a yardstick vertically 1 foot from the edge of the desk with the bottom of the yardstick level with the desk. The yardstick should be dropped without any signals as to when it will be dropped. The first student should raise his or her hand from the desk and catch the falling yardstick as quickly as possible. Turn the yardstick so that the number of inches the stick has fallen can be read above the students' hand (see Figure 3–1.) Determine the reaction time in seconds from the Reaction Time chart. Determine the consequences of the student's reaction time from the Consequences of Reaction Time chart.

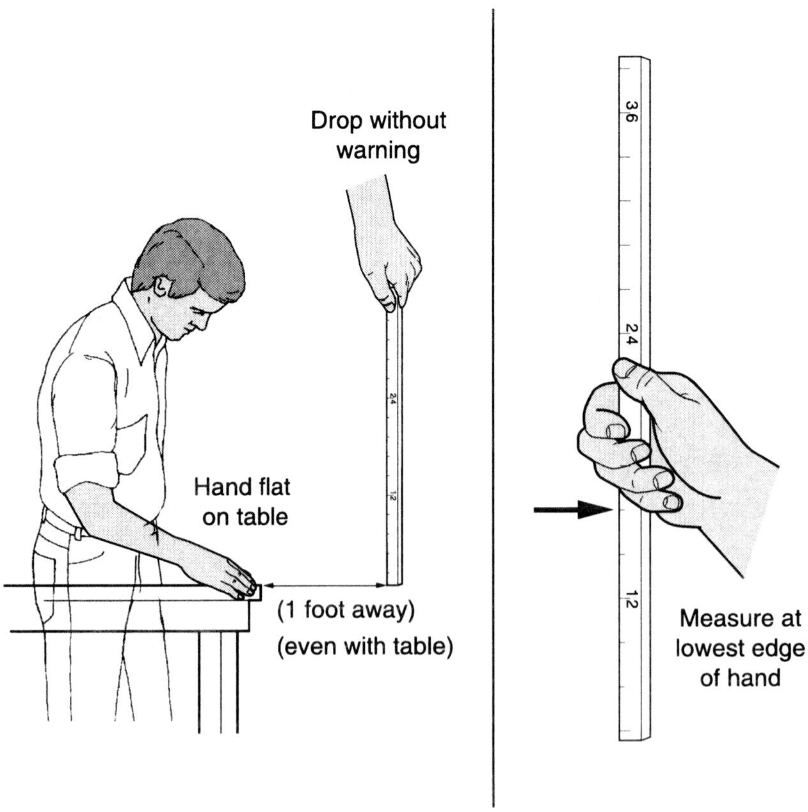

Figure 3–1

Hypothesis

Write a brief statement guessing if you think your reaction time will be faster, slower, or average than the times of others in your class. Use the Reaction Time chart below as a gauge (1–4 being fast, 8–10 average, and 16+ slow).

____

____

____

Observations

Collect Data

Reaction Time

Distance in Inches	Reaction Time in Seconds	Distance in Inches	Reaction Time in Seconds
1	0.072	19	0.315
2	0.102	20	0.323
3	0.125	21	0.331
4	0.144	22	0.338
5	0.161	23	0.346
6	0.177	24	0.354
7	0.191	25	0.361
8	0.204	26	0.368
9	0.216	27	0.375
10	0.228	28	0.382
11	0.239	29	0.389
12	0.250	30	0.395
13	0.260	31	0.402
14	0.270	32	0.408
15	0.279	33	0.415
16	0.288	34	0.421
17	0.298	35	0.427
18	0.306	36	0.433

Determine the student's reaction time in the yardstick drop exercise, then determine what the following pieces of equipment can do in that same length of time. For example, if a student's reaction time is 0.3 seconds, a rock could be thrown from a lawnmower 75 feet or an arm could be wrapped five times around a PTO (power takeoff) shaft at 1,000 rpm before the student could react to an emergency situation.

Consequences of Reaction Time

Reaction Time in Seconds	PTO @ 540 rpm (rotations)	Rotations of a Pulley	Tractor @ 20 mph (distance in feet)	PTO @ 1,000 rpm (rotations)	Object Thrown by Mower (distance in feet)
0.1	0.9	1.2	2.93	1.67	25
0.2	1.8	2.4	5.86	3.34	50
0.3	2.7	3.6	8.79	5	75
0.4	3.6	4.8	11.72	6.68	100
0.5	4.5	6	14.66	8.35	125

Conclusions

Compare your time with other students in your class. Did this data support or reject your hypothesis?

____

____

EXERCISE 2 THE AXE MEN/WOMEN GAMES

Background

As described in the text, safety in the forest begins with proper safety clothing and equipment for the workers. Clothing and equipment should be designed to protect the workers from the hazards of falling debris, misdirected cutting tools, and crushing by heavy objects. In this lab, you will become familiar with the proper safety gear needed while working in the forestry industry.

Materials Needed

Two or more of each of the following:

- Steel-reinforced boots
- Safety glasses
- Hard hats
- Gloves
- Long-sleeved work shirts
- Chaps
- Durable work pants
- Ear protection
- Axe
- File

Procedures

The instructor will demonstrate the proper and safe way to sharpen an axe using a file. The class will then be split up into teams that will compete against each other for the title of "The Ultimate Axe Man (or Woman) Champion." The classroom will be the venue for the Axe Men/Women Games. On one side of the course, all of the forestry safety equipment will be laid out in the order mentioned above. The teams will be at the starting line on the opposite end of the classroom. This competition will be a relay race, with a purpose.

When the race begins, the first two competitors will race to the equipment, put on the steel-toed boots and shout, "I have on my steel-reinforced boots." The racers will then run back to their teams, with the boots still on and tag the next person in line. The first person will give the boots to the second person. The second person will put on the boots and run to the equipment. He or she will then put on the next item, the safety glasses. The racer will then shout, "I have on my steel-reinforced boots and my safety glasses," before running back to his or her team and transferring the equipment to the next person. This will continue until the last person on the team is fully dressed in all of the safety equipment and has verbally identified all of the safety equipment he or she is wearing. This team member will then move to the axe sharpening station and sharpen an axe to the satisfaction of the instructor. The first team to complete the challenge wins bragging rights and the title of "The Ultimate Ax Men or Women."

EXERCISE 3 COLOR CODING

Background

A large part of accident prevention is knowing where the danger areas are. In all operations, there are certain areas that are more hazardous than others. These areas should be color-coded to represent the type and degree of danger. Colors are easily recognized at a glance and can be very useful in alerting workers to dangers. However, color coding should always be used in combination with clearly printed labels in case any worker has difficulty distinguishing colors. The following are some of the colors used to indicate the need for safety precautions:

Red: For fire apparatus and equipment, safety containers for flammable materials, and emergency devices like switches for emergency stopping of machinery, stop bars, and so on.
Orange: For potentially dangerous parts of machinery or equipment that may cut, crush, shock, or otherwise injure a person.
Yellow: The color of caution; for physical dangers, such as slipping, tripping, falling, and caught-between or striking-against hazards.
Green: The color of safety; for first-aid equipment locations.
Blue: A caution color; for warning against the use or movement of equipment being repaired or worked on or against the starting of equipment.
Purple: The basic color for warning of radiation hazards.
Black, white, or a combination: For designating and helping to control traffic movement and for marking aisles, housekeeping areas, and similar areas.

Procedures

If your school has a mechanics laboratory, go through the lab and identify areas that should be color-coded. An alternative might be to visit another mechanics shop not associated with your school or an area where forest products are harvested or processed. If the hazardous areas are not color-coded, then secure permission and paint the area using the proper color coding. The following chart can be used for your safety inventory.

Location of Hazard	Specific Hazard	Proper Color Code	Color Coding Needed?

Chapter 4

SILVICULTURE—*WHAT THE HECK IS SILVICULTURE?*

OBJECTIVES

After completing these lab exercises, students should be able to:

- explain how seeds are protected by pine trees
- discuss how pine seeds are dispersed
- develop a hypothesis
- explain the process of germination
- explain scarification of seeds

EXERCISE 1 COME OUT, COME OUT, WHEREVER YOU ARE—THE PINE CONE

Background

In nature, trees reproduce by forming and dispersing seed. A tree flowers, produces a seed, expels the seed, and a new tree germinates from the seed. Nature has several ways in which to protect seeds and to make sure they are dispersed. In this exercise, you will have the opportunity to observe how a pine tree protects and disperses its seed.

Materials Needed

- Pine cones
- Large pan

Procedures

Gather several pine cones from underneath a pine tree. In the lab, soak the cones in a pan of water until they are thoroughly wet. Remove them from the water and make notes on any changes in the cones. Within a few minutes after soaking, the cones should close up tightly. Allow the cones to dry and observe the changes in the cones. As they lose moisture, the cones should begin to open. Once they are completely dry, the cones should be fully open.

Hypothesis

In the woods, pine cones get wet and they dry out. As observed from your experiment, the cones open and close depending on the moisture content. This phenomenon must serve a purpose. When a scientist observes such a phenomenon, he or she develops a hypothesis. A **hypothesis** is an untested, logical explanation of an observed phenomenon. The scientist makes his or her best guess about why the event or phenomenon occurred and then designs an experiment to prove or disprove that hypothesis. Try to develop your own hypothesis about why the cones close in wet weather and open in dry weather. Look for hints, such as the fact that seeds are developed and stored inside the sections that open and close.

Hypothesis: ____

____

____

Observations

Collect Data

Time	Observed Changes (include drawings)
Time in water:	
Time it took to dry:	

Conclusions

Review your hypothesis. Do your observations support or reject it? Explain your answer.

____

____

EXERCISE 2 COME ON IN, THE SOIL IS FINE—SEED SCARIFICATION

Background

Almost all trees that are used commercially begin life as seeds. The seed contains all of the genetic material that determines the characteristics of the tree. When the tree produces the seed, the seed is usually in the dormant stage. This is to prevent the seed from germinating too quickly or at the wrong time of the year. For the seeds to germinate, they have to have the proper temperature and moisture level and the correct amounts of light and oxygen. The proper mix of these conditions vary considerably with various seeds.

Seeds generally have to undergo a period of cold weather before they will germinate, although some may simply require a certain amount of time between maturation and germination. The period of cold or the interval of time allows the seed to survive the winter.

Some seeds have very hard seed coats that resist the entrance of water. This is particularly true of tree seeds, which need a very hard seed coat to survive in the forest, where they may be subjected to freezing, flooding, or fire. The seed coat has to be weathered or eroded away before moisture can enter the seed and the germination process can begin. This process is known as **scarification**.

In nature, scarification takes place over a period of time. For example, a fire may soften the seed coat enough for the seed to germinate, or the weather may wear away the seed coat. In commercial nurseries, seeds may be scarified by artificial means, such as by using mechanical abrasion, by soaking in water for several hours, or by subjecting the seeds to acid. This can be a tricky process because the amount of scarification has to be just right. If it is too little, moisture cannot enter the plant; but if it is too much, the embryo in the seed will be damaged and the seed will not germinate.

In this exercise, you will experiment with various methods of scarifying seeds. At the end of the exercise, you will draw a conclusion about which method is best.

Materials Needed

- Tree seed or other seed with a very hard coat
- A roll of paper towels
- An oven
- A flat pan
- Sandpaper
- A large bowl of warm water
- Rubber bands
- Large, resealable plastic bags

Procedures

1. Count out several groups of good seeds, with 100 seeds in each group. Submit the groups to the following treatments.
2. Place a group of seeds in a bowl of warm water overnight.

3. Place the seeds on a flat pan and heat it in the oven at 1500°F for 1 hour. Vary the temperature and time for the other groups of seeds.
4. Lay a sheet of sandpaper (about 100 grit) on a table with the sand side up. Lay a group of seeds flat on the sandpaper and lay another sheet of sandpaper on top of the seeds, with the sand side facing the seeds. Without applying too much pressure, gently roll the seeds between the sheets of sandpaper (see Figure 4–1). You might want to use varying pressures and different sanding times for different groups of seeds. Think of other ways to scarify the seed.

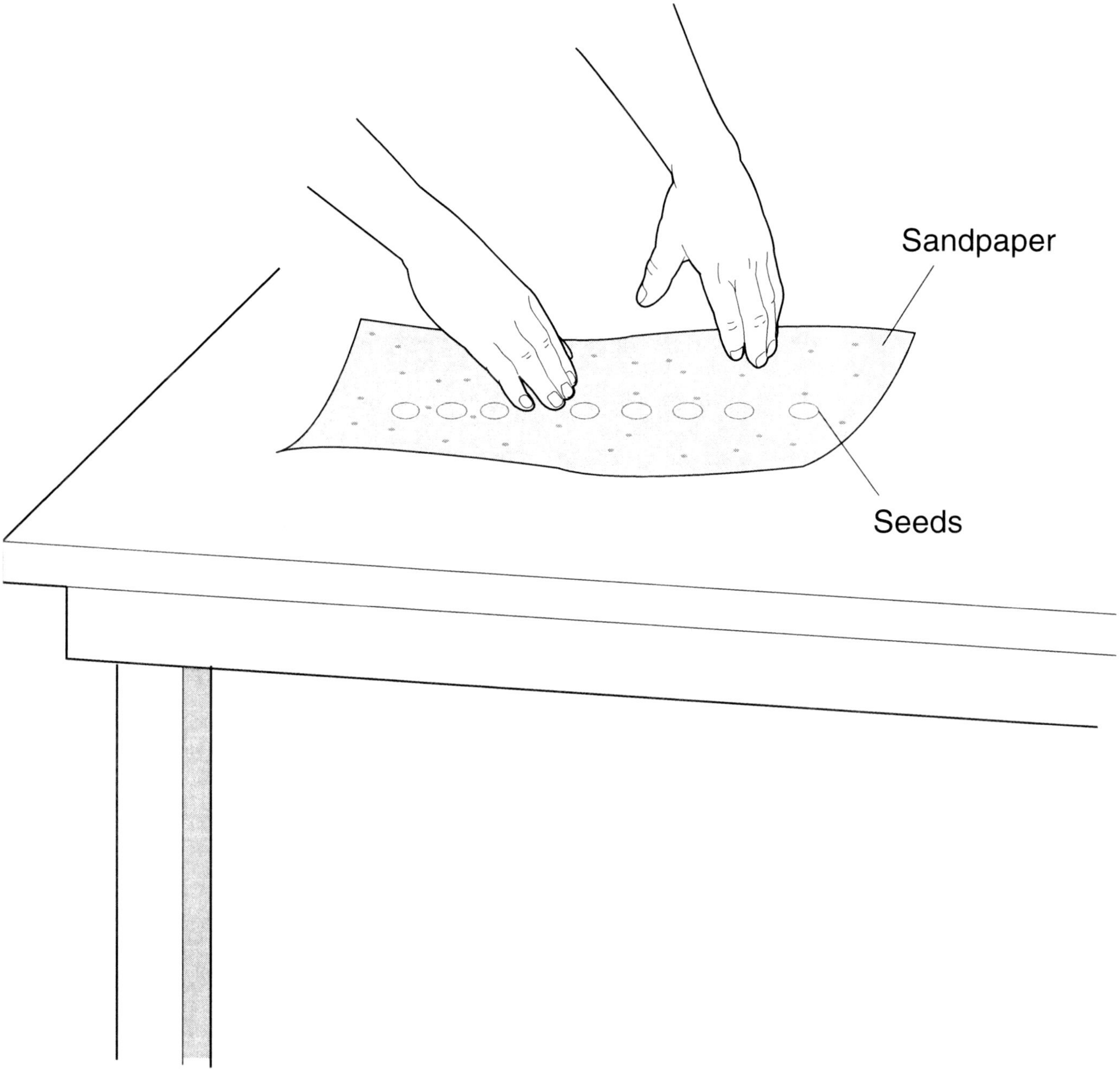

Figure 4–1

5. As soon as each group of seeds receives a treatment, roll out about 2 feet of paper towels and place the treated seeds in 10 rows of 10 seeds. Start each row about 1 inch from each side of the towel. Roll the towel up, keeping the seeds in place in their rows. Twist the ends of the roll and secure it with rubber bands (see Figure 4–2). Wet the roll and place it in a large plastic bag. Be sure to label the bag regarding the treatment the seeds received. Also make a roll of seeds that have received no treatment.

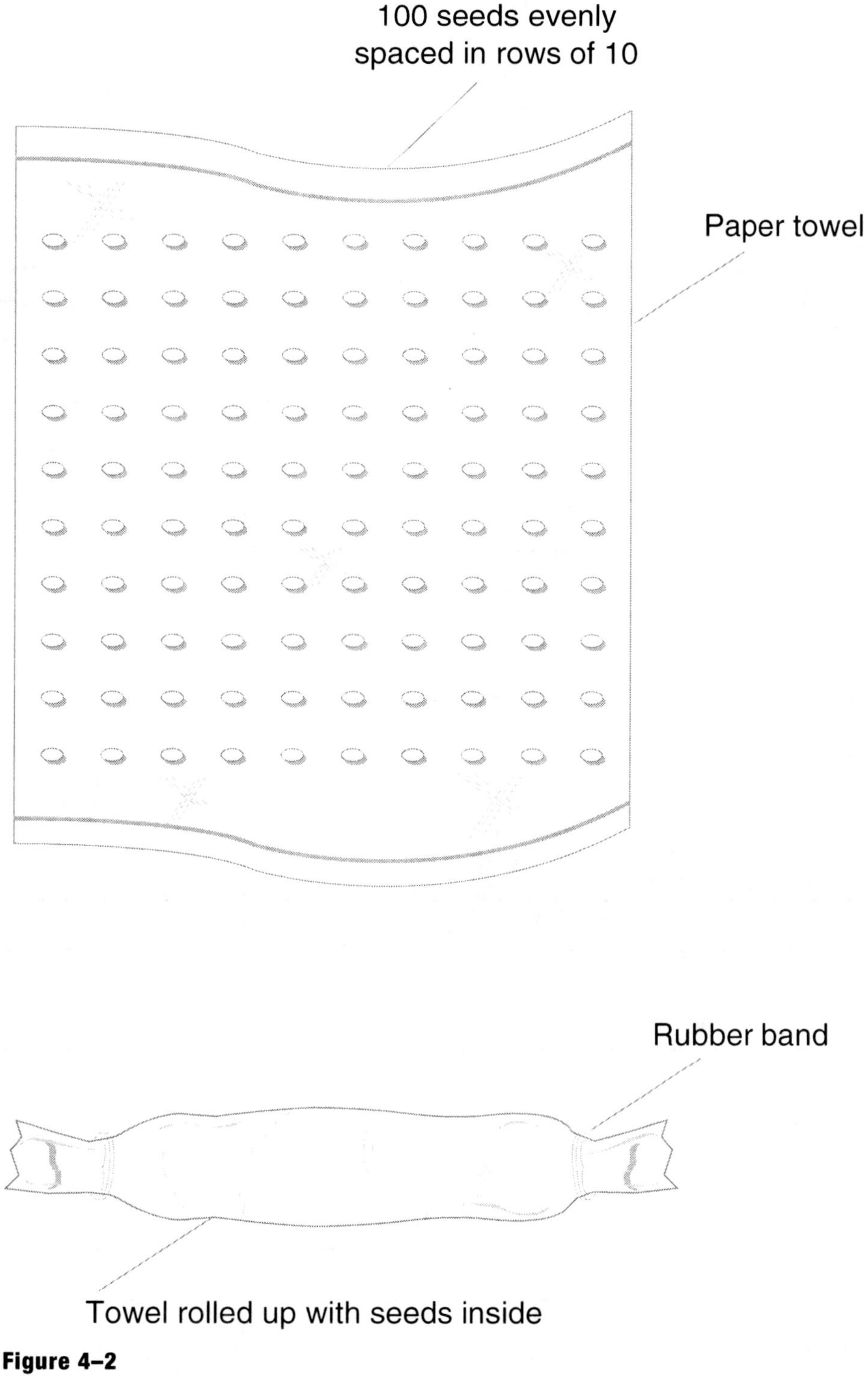

Figure 4–2

6. Place all the bags in a warm dark place. Each day, open each bag and roll out the paper towel. Count the seeds that have germinated, calculate the percentage of germination, and record the results in the following chart. All of the seeds should germinate within 10 days to 2 weeks.

Hypothesis

Before you begin your experimentation, write a statement predicting which group of seeds you think will germinate first.

____

____

Observations

Collect Data

Using the chart below, keep a record of the treatments and then the resulting changes you observed.

Treatment	Day Number	Percentage of Germination

Treatment	Day Number	Percentage of Germination

Conclusions

Were you right? Comment on whether your observations supported or rejected your hypothesis. Explain in the space provided.

Chapter 5

MEASUREMENT OF FOREST RESOURCES—*ACCORDING TO MY CALCULATIONS . . .*

OBJECTIVES

After completing these lab exercises, students should be able to:

- locate townships using a written description
- identify on a map a certain parcel of land using a description from a deed
- determine the number of acres in a plot of ground using a deed description
- estimate land areas
- estimate the amount of timber on a plot of land
- determine the value of standing timber

EXERCISE 1 LOCATING LAND PLOTS—DOES IPHONE HAVE AN APP FOR THIS?

Background

Foresters need to accurately locate tracts of land to make sure they are working in the correct location. A surveying system has been developed to help in the location of tracts. Using a land deed description, a map, and a beginning reference point called an **initial point**, foresters can locate the boundaries of any parcel of land in the United States.

Large tracts of land are divided into squares, each consisting of 36 square miles of land. These 36-square-mile divisions are called **townships**. A township is located by moving to the north or south of a horizontal east-west running line, called a **baseline**, and then going east or west of a vertical north-south running line, called a **principal meridian**. A township is identified using consecutive numbers beginning at the baseline and then moving north or south. For example, the third southern township located 3 grids south of the baseline is described as T3S (T = township, # grid lines, S = south). The **range** (R) refers to the township being to the east or the west of the principal meridian. A township located 3 grids to the west of the principal meridian is described as R3W (R = range, # grids, W = west). So, the location of a township that is 3 grids south and 3 grids west is called T3S, R3W.

Procedures

Using the map in Figure 5–1, locate the following townships, then shade in those squares and label them.

a. T4N, R1E

b. T3S, R4W

c. T2N, R2E

d. T1N, R3E

e. T4S, R2E

f. T3N, R3W

g. T1N, R3W

h. T2S, R4W

i. T1S, R3W

j. T4S, R3E

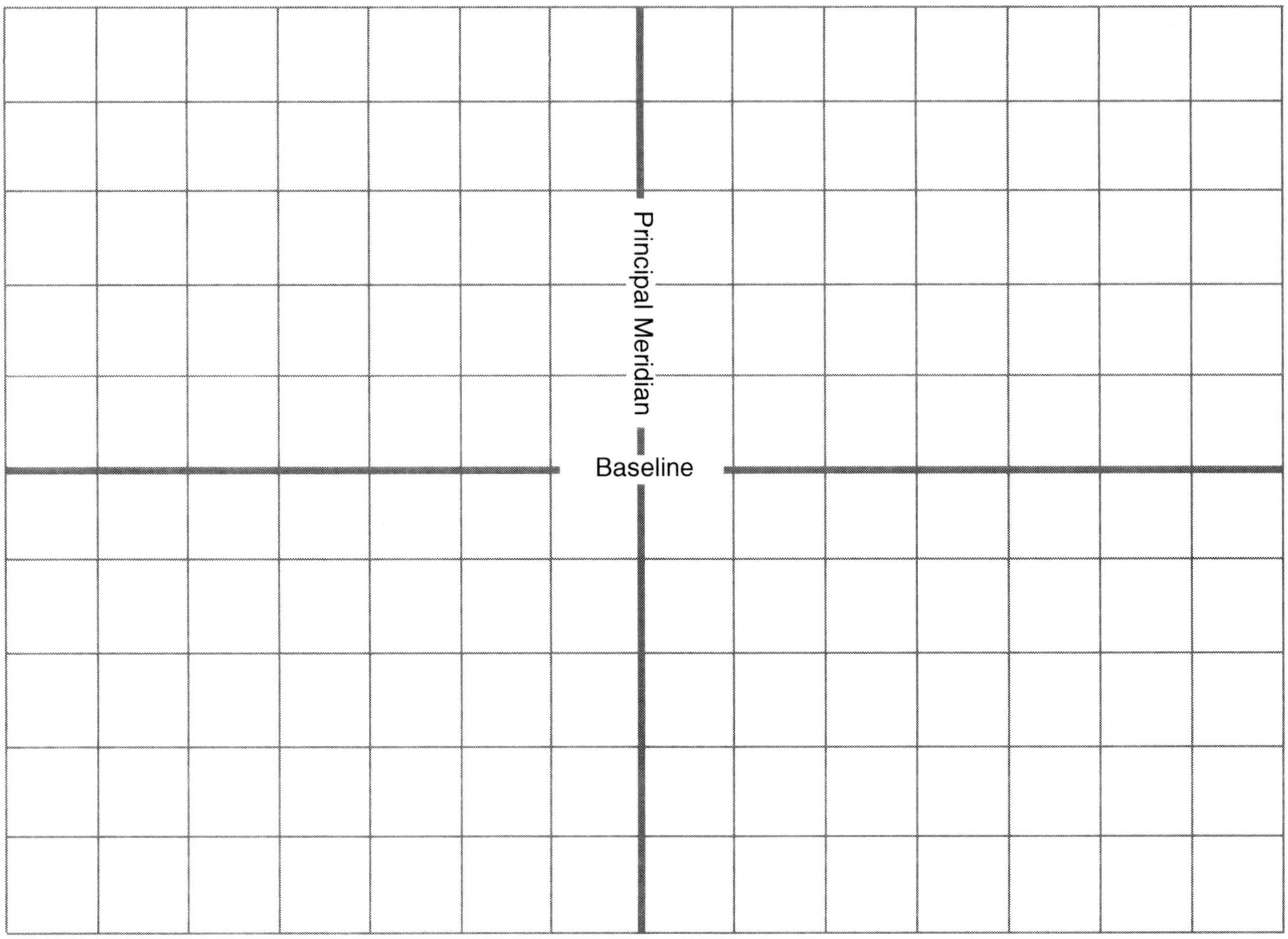

Figure 5–1

Each township is divided into 36 squares, called **sections**. Each section is numbered consecutively, beginning in the northeast corner, as shown in the following diagram. Each section measures 1 square mile and contains 640 acres.

Numbering System for Sections

6	5	4	3	2	1
7	8	9	10	11	12
18	17	16	15	14	13
19	20	21	22	23	24
30	29	28	27	26	25
31	32	33	34	35	36

Sections are divided further into **quarter sections**, each of which contains 160 acres. Quarter sections are then divided into four parts called **forties**, because they each contain slightly fewer than 40 acres. The forties may be further divided and described as **quarters of a forty** or **portions of the forty**. A deed description may read: "A parcel of land located in the NW¼ of the SE¼ of the SW¼ of section 23 T2NW, R3SE." This land area would contain 10 acres and would be located in the shaded area of the township (T2NW, R3SE) shown in the following diagram. Using the diagram, locate, shade in, and label the following land descriptions. Below each description, write the number of acres in the description. Remember to work backward in dividing each section into quarter sections and forties.

a. SW¼ of the NW¼ of the SE¼ of the NW¼ of section 18 T2NW, R3SE.

Number of acres: ________

b. SE¼ of the SW¼ of the NW¼ of the NW¼ of section 3 T2NW, R3SE.

Number of acres: ________

c. NE¼ of the NE¼ of the NW¼ of section 30 T2NW, R3SE.

Number of acres: ________

d. S½ of SE¼ of section 34 T2NW, R3SE.

Number of acres: ________

e. NW¼ of the NW¼ of section 20 T2NW, R3SE.

Number of acres: ________

f. SE¼ of section 7 T2NW, R3SE.

Number of acres: ________

g. SW¼ of the NW¼ of the NE¼ of section 23 T2NW, R3SE.

Number of acres: ________

Township T2NW, R3SE

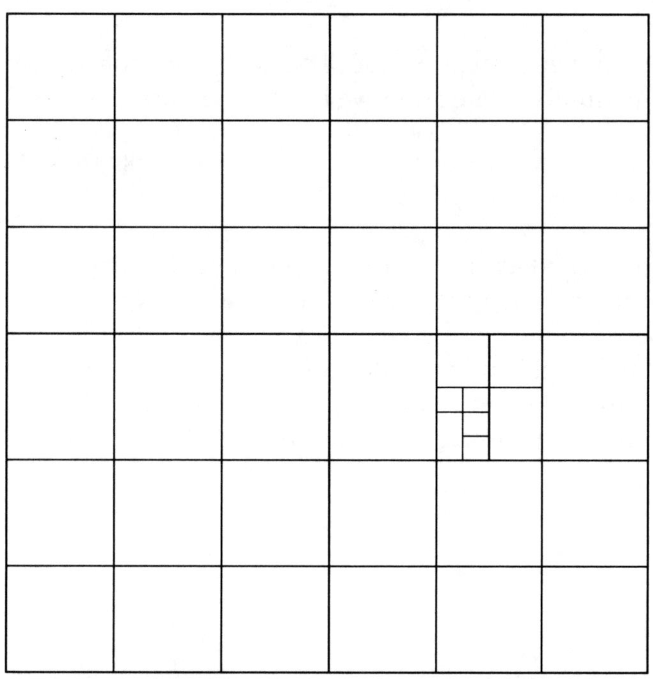

EXERCISE 2 STEP ON THE SCALE—ESTIMATING TIMBER VOLUME

Background

Foresters must accurately estimate how much volume of timber is on a plot of ground by determining the amount on a smaller plot of ground that is representative of the larger plot and using that amount to figure the estimate for the entire acreage.

Materials Needed

- 100-foot measuring tape (or measuring chain)
- A log-scale stick (Doyle, Scribner, or International)
- A Biltmore stick or an inclinometer
- A diameter tape or calipers
- Several wooden stakes

Procedures

Step 1

Locate a large open area on the school campus and drive a stake into the ground. Using the measuring tape, measure off 100 feet and drive another stake into the ground. Using normal steps, count the number of paces it takes you to walk between the two stakes. Do this five times and record your findings in the space provided. After you have recorded all your paces, add all the steps together and divide by 5. This will give you the average number of steps in 100 feet.

First try: _________ paces
Second try: _________ paces
Third try: _________ paces
Fourth try: _________ paces
Fifth try: _________ paces
Total number of paces: _________ ÷ 5 = _________ average number of steps in 100 feet

Next, divide the average number of steps by 100. This will give you the average number of linear feet in each of your steps. Record the number here and save it for future use:

_________ = average number of linear feet/step

Step 2

Locate a number of trees in a wooded area nearby. Drive a stake into the ground at a beginning point and determine the shape of the wooded area. If the trees are in a square or rectangle, walk the length of one side, counting your paces as you go, and then drive a stake into the ground at the corner. Turn at a 90-degree angle and walk the length on that side, counting your paces as you go. If the plot is a circle, walk the perimeter of the trees and count your steps. If the plot is a triangle, walk the longest length of the triangle (the base) and the distance from the base to the highest point. Do not forget to count your steps and record them.

For each side of the plot, convert the number of your steps into linear feet. Use the number you recorded earlier as the number of feet in one of your steps. Simply multiply the number of steps by the average number of feet in one step. This will give you the approximate dimensions of the plot.

Step 3

Next, you need to figure how many square feet are in the plot. Use the following figures to determine the area of the plot:

- **Square or rectangle:** Multiply length by width.
- **Triangle:** Multiply one-half the base times the height of the triangle.
- **Circle:** Divide the number of feet in the circumference (the distance around the plot) by 3.17 (π). This will give you the diameter of the circle. Divide the diameter by 2 and you will have the radius of the circle. Use the radius in the formula πr^2 to figure the area of the circle.

Step 4

Once you have the number of square feet in the area, you can determine the number (or fraction of) acres in the plot. To do this, simply divide the number of square feet in the plot by 43,560 (the number of square feet in an acre). Record your finding here for later use.

_________ = acres in plot

Step 5

The next step is to determine the volume of logs in the plot. To do this you must first find the merchantable height and **dbh** (diameter at breast height) of each tree. The dbh can be measured in two ways. One is to use a diameter tape. This measuring tape is different from a regular measuring tape in that it measures a tree's diameter rather than the circumference. Place the tape around the tree at breast height and read the diameter at the scale where the end of the tape meets the rest of the tape.

Step 6

Another method is to use a cruising stick such as a Biltmore stick. Make sure you hold the stick level at a distance of about 25 inches from your eye and 4½ feet from the ground. Close one eye and, without moving your head, read the diameter of the tree from the scale on the stick (see Figure 5–2). Be sure to record all your data.

Step 7

To determine the height of the tree, use either an inclinometer or a cruising stick. The **inclinometer** is a handheld instrument that is sighted through. A scale in the line of sight determines the height of the tree. Another common method is using a **cruising stick**. A distance of 66 feet is paced off from the tree, and the stick is held vertically at eye level. Sight the lower end of the stick on the bottom of the tree where it would be cut (stump height). Without moving your head, read the scale on the stick where the tree measures about 5 inches in diameter. (See Figure 5–3.) Read the scale to the nearest one-half log. Logs are measured in 16-foot lengths.

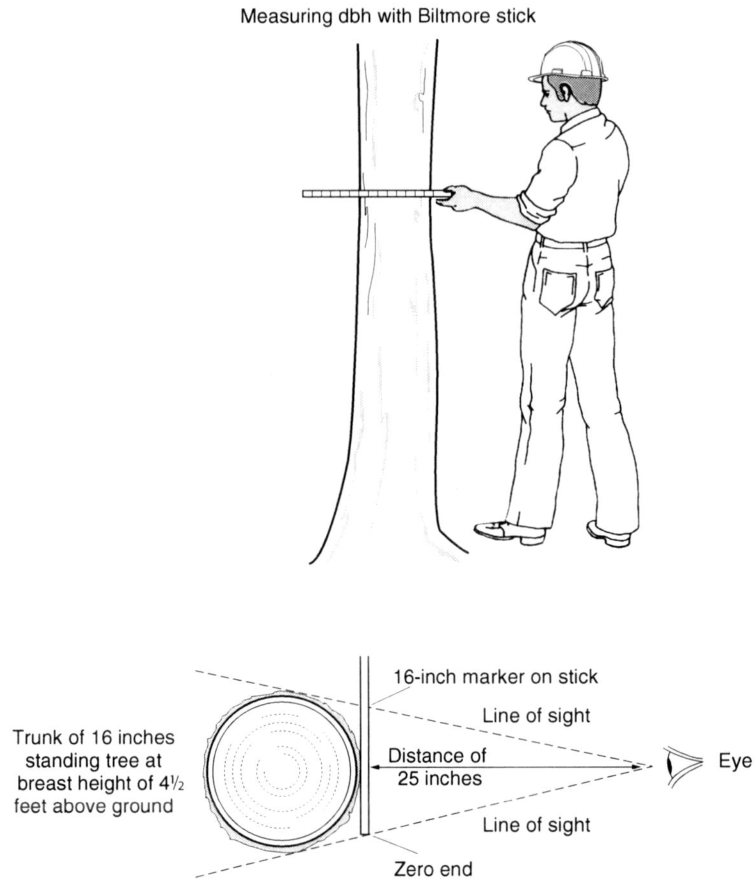

Figure 5–2

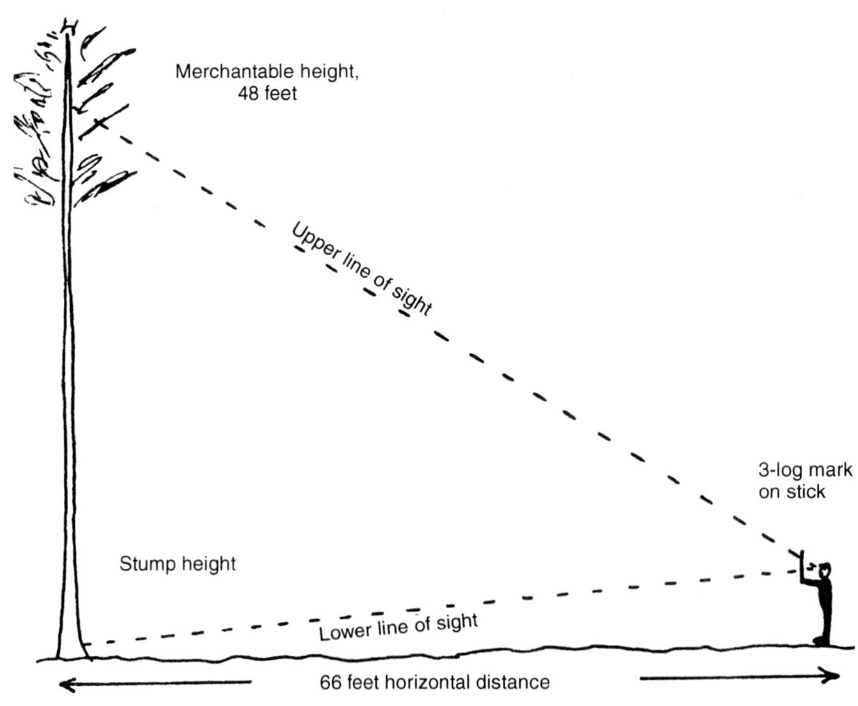

Figure 5–3 *(Courtesy Georgia Department of Education)*

Record all your measurements in the following table:

Observations

	Specie (oak, pine, and so on)	Diameter	Height	Volume
Tree #1				
Tree #2				
Tree #3				
Tree #4				
Tree #5				
Tree #6				
Tree #7				
Tree #8				
Tree #9				
Tree #10				

Step 8

The volume of timber in board feet can be determined by using a volume table. Foresters use several tables. The one provided on page 40 is a very common table known as the Doyle scale. Use this table to determine the volume of timber in each tree you measured.

When you get all of the measurements filled in on the table and all of the volumes entered, add all the volumes together and determine how many board feet of lumber you have for the plot of ground you measured. Figure how much volume would be on an acre of land with trees like the ones you measured. For example, if your plot measured ½ acre, multiply the entire volume by 2 to obtain a volume for an acre. Go to the Internet and locate prices for the type of wood on your plot and calculate how much an acre of timber like your plot would be worth. Make sure your prices are for logs and not for finished lumber.

Conclusions

Think back to the calculations you just did. Many of these measurements were "eye balled" or estimated. Do you think that this method is the best way to determine lumber volume? Explain your answer.

____

____

____

____

____

____

Gross Volume of Tree, Doyle Log Rule Volume (board feet) by Number of Usable 16-Foot Logs

dbh	1 log	1½ logs	2 logs	2½ logs	3 logs	3½ logs	4 logs	4½ logs	5 logs	5½ logs	6 logs
10	14	17	20	21	22						
11	22	27	32	35	38						
12	29	36	43	48	53	54	56				
13	38	48	59	66	73	76	80				
14	48	62	75	84	93	98	103				
15	60	78	96	108	121	128	136				
16	72	94	116	132	149	160	170				
17	86	113	140	161	182	196	209				
18	100	132	164	190	215	232	248				
19	118	156	194	225	256	276	297				
20	135	180	225	261	297	322	346	364	383		
21	154	207	260	302	344	374	404	428	452		
22	174	234	295	344	392	427	462	492	521		
23	195	264	332	388	444	483	522	558	594		
24	216	293	370	433	496	539	582	625	668		
25	241	328	414	486	558	609	660	709	758		
26	266	362	459	539	619	678	737	793	849		
27	292	398	505	594	684	749	814	877	940		
28	317	434	551	650	750	820	890	961	1,032	1,096	1,161
29	346	475	604	714	824	902	980	1,061	1,142	1,218	1,294
30	376	517	658	778	898	984	1,069	1,160	1,251	1,339	1,427
31	408	562	717	850	983	1,080	1,176	1,273	1,370	1,470	1,570
32	441	608	776	922	1,068	1,176	1,283	1,386	1,488	1,600	1,712
33	474	654	835	994	1,152	1,268	1,385	1,497	1,609	1,734	1,858
34	506	700	894	1,064	1,235	1,361	1,487	1,608	1,730	1,866	2,003
35	544	754	964	1,149	1,334	1,472	1,610	1,743	1,876	2,020	2,163
36	581	808	1,035	1,234	1,434	1,583	1,732	1,878	2,023	2,173	2,323
37	618	860	1,102	1,318	1,534	1,694	1,854	2,013	2,172	2,332	2,492
38	655	912	1,170	1,402	1,635	1,805	1,975	2, 148	2,322	2,491	2,660
39	698	974	1,250	1,498	1,746	1,932	2,118	2,298	2,479	2,662	2,844
40	740	1,035	1,330	1,594	1,858	2,059	2,260	2,448	2,636	2,832	3,027

EXERCISE 3 CORDWOOD CALCULATIONS

Background

Many people sell and buy firewood for heating their homes, to take camping, or just to use in outside fire pits and BBQs. Most firewood is sold by the cord. A **standard cord** is a stack of wood 4 feet × 4 feet × 8 feet and is 128 cubic feet. It is important that both buyers and sellers know how to measure the volume of wood in a cord to ensure fair and honest commerce.

Procedures

Step 1

If cut and split firewood is available, have the students stack it in into a neat pile of any dimension. Students should measure the height, width, and length of this pile of wood.

List the height (h) ________
List the width (w) ________
List the length (l) ________

Multiply these three numbers together:

(h) ________ × (w) ________ × (l) ________ = ________cubic feet

Step 2

Dividing the number of cubic feet by 128 will give you the number of cords you have in the pile. It may be more than 1 cord or it may be a fraction of 1 cord. Do the calculation below:

No. of cubic feet ________ ÷ 128 = ________ cords

Step 3

If this firewood is selling for $100 per cord, how much is the cost per cubic foot?

$100 ÷ 128 cubic feet per cord = $________/cubic foot

How much should your pile of wood sell for at this rate? Do the calculation below:

No. of cubic feet of wood in your pile ________ × $________/cubic foot = $________ selling price

Step 4

With your newly mastered cord calculating skills, give this next example a try.

Example

Imagine that you are working as an intern for a state park campground. One of your daily duties is to distribute firewood to each camp site. The campground replenishes its supplies once a month (30 days). You will need to know how much wood the campground will need to purchase to keep the park's visitors happy campers.

There are 20 sites in the park. If every site is reserved and is allotted three, 4-inch × 4-inch × 24-inch logs a day, how many cords of wood will you need to order? How many cubic feet is this? If the firewood is $100 per cord, what will a month's supply cost?

Calculation

(Hint: You will need to convert cubic feet into cubic inches. Be careful, it is not as simple as 128 cf. × 12 in.)

Chapter 6

HARVEST AND REFORESTATION PRACTICES—*SOW WHAT YOU REAP*

OBJECTIVES

After completing these lab exercises, students should be able to:

- determine factors that must be considered in planning logging operations
- determine the slope of a tract of land
- construct a plan for logging a specific area

BACKGROUND

To make use of the nation's forest resources, trees must be harvested and removed from the forest. At the very least, the ecology and the environment will be disturbed. There is probably no method that will work best in all areas under all conditions. The terrain, the weather, environmental conditions, and the attitudes of the people in the area are different in each location. In addition, the type and age of the forest stand can have a large effect on the type of logging operation used. Each situation must be analyzed, and decisions must be made based on a wide variety of information and options.

EXERCISE 1 SOMETHING IS A LITTLE OFF—DETERMINING SLOPE

Background

The **slope** of an area is an important factor to understand for those who work in the forest. Slope determines what procedures will be used to ensure a safe logging operation. A critical part of planning any logging operation is determining the steepness of the slopes from which the timber must be harvested. This determination can be made by using simple tools.

Materials Needed

- A straight 2 × 4 that is 10 feet long
- Spirit level
- Measuring tape

Procedures

Find a wooded area that has trees of harvestable size. Go to the highest point on the tract of land and place one end of a 10-foot 2 × 4 on the highest point with the narrow side up. Place a spirit level on top of the 2 × 4 and raise the end of the board until it is level. Measure the distance from the top of the board to the ground and record your measurement. Mark the place where the ruler touched the ground and move the end of the board to that spot. Repeat the steps you just completed and again record the distance.

Continue "stepping" your way down the slope until you have repeated the exercise 10 times. This will give you 100 feet of horizontal distance down the slope. By adding all the recorded distances from the top of the level board to the ground, you will calculate the vertical drop in 100 feet. If you divide the vertical drop (in feet) by 100 feet, you will calculate the percentage of slope (see Figure 6–1).

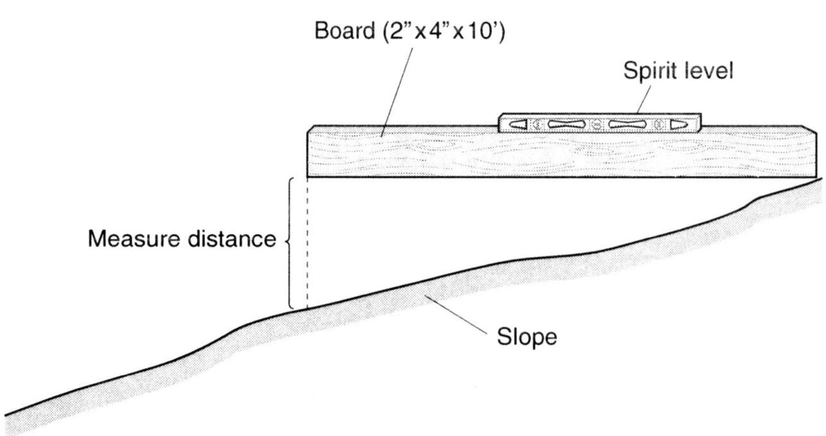

Figure 6–1

Observations

Collect Data

	Drop in Feet
1st	
2nd	
3rd	
4th	
5th	
6th	
7th	
8th	
9th	
10th	

Research

Obtain a copy of the "Best Management Practices" for your state. These guidelines can generally be obtained from the department of forestry in your state, or they can be found on the Internet. You might want to compare the guidelines for your state with those of other states. List the regulations regarding the slope of the forest for the following:

Area	Regulations or Restrictions (due to an area's slope)
Riparian zones	
Logging roads	
Harvest techniques	
Seeding method	
Type of machinery used	

Conclusions

Develop a logging plan using the "Best Management Practices" for the wooded area where you measured the slope. If possible, have a forester come to class and evaluate your plan.

EXERCISE 2 I'M SEEING DOUBLE—DETERMINING BASAL AREA

Background

If a forester is contemplating the harvest of a plot of timber, he or she may wish to know how much timber is on the land and whether the plot needs to be thinned. In Chapter 5, you learned a method of cruising timber using a Biltmore stick. In this exercise, you will learn a method called **basal area**. This method measures the square footage of timber based on the area of the cross sections of trees at diameter breast high (**dbh**).

One way to measure basal area is to use a prism held to the eye. These prisms come in different "factors," the most common being a 10 factor. To use the prism, the forester stands in the middle of a stand of trees and holds the prism to his or her eye. Each tree is examined through the prism. Through the prism, a chunk of the tree trunk appears to be misaligned with the rest of the tree. Those chunks that appear to be within the vertical lines of the tree are counted; those that are outside the lines are not counted. Some trees may appear to be borderline; every other borderline tree is counted (see Figure 6–2).

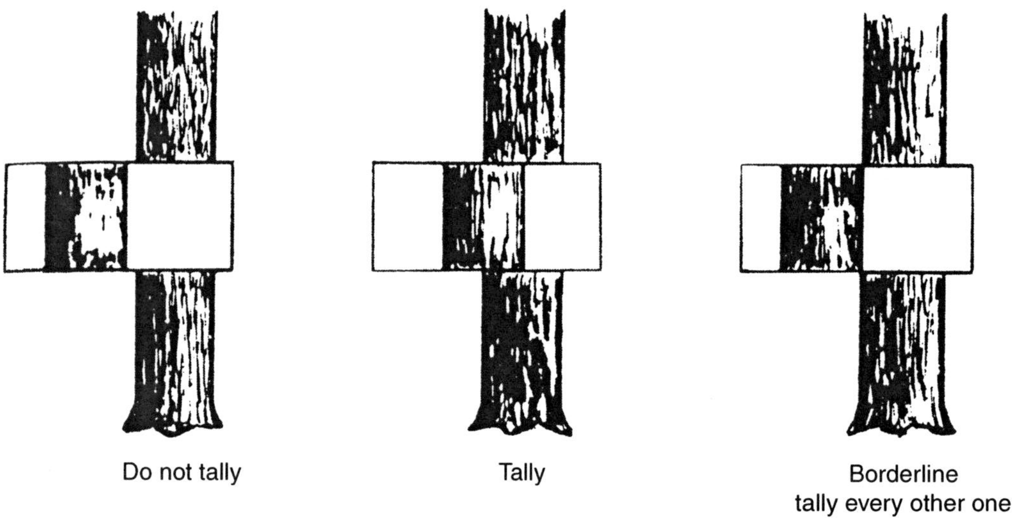

Figure 6–2 *(Courtesy Georgia Department of Education)*

Another way to measure basal area is to use a device called a **tube angle gauge** that you can make yourself using PVC pipe.

Materials Needed

- Piece of 1-inch inside diameter PVC pipe *exactly* 33 inches long, including the cap at the end
- End cap for 1-inch PVC pipe
- Drill with a •-inch bit

Procedures

First drill a •-inch peephole in the exact center of the pipe cap. Place the cap on the end of the pipe. You may want to glue it in place, but as long as the cap stays in place, glue is not necessary. Carefully cut the pipe to exactly 33 inches including the cap (see Figure 6–3).

Locate a group of trees you wish to measure. The size of the plot does not matter. Place the cap with the peephole up to your eye, pointing the tube at dbh on the tree. If the trunk of the tree at dbh appears larger than the hole at the end of the tube, the tree should be tallied. Trees that appear smaller than the hole do not count (see Figure 6–4).

All trees visible from the sampling point must be counted or rejected. The count of trees multiplied by 10 gives an estimate of basal area per acre. In the diagram in Figure 6–4, only two trees are counted, so the basal area estimate is 20 square feet per acre.

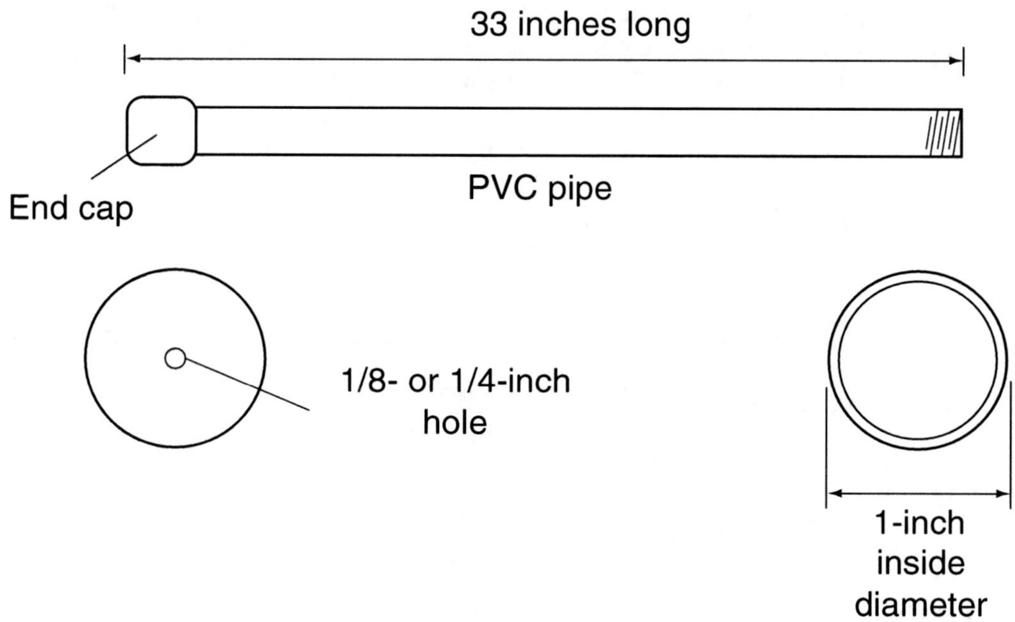

Figure 6–3 *(Courtesy Georgia Department of Education)*

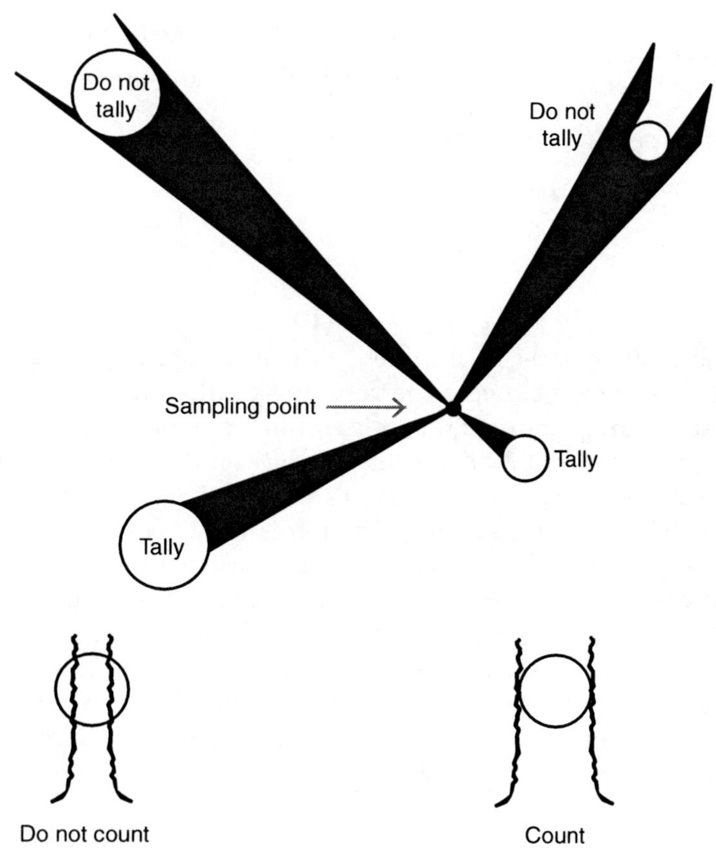

Figure 6–4 *(Courtesy Georgia Department of Education)*

Reliable estimates require more than one sample, of course. It is recommended that on areas under 50 acres, sampling points should be on a three-chain grid. On larger areas, the cruiser can go to a four- or five-chain grid in laying out sampling points. The cruiser must then visit each sampling point, look in every direction through the instrument, and count the number of trees that appear to have a dbh larger than the crosspiece. The eyepiece of the angle gauge should pivot on the sampling point until the count is completed, except that it may be temporarily moved sideways perpendicular to the line of sight to clear nearby brush or trees likely to mask or block other qualifying trees.

Example

Suppose that the cruiser has tallied a total of 240 qualifying trees at 30 sampling points on an area.

$$\begin{aligned}\text{Estimated basal area per acre} &= 10 \times \frac{\text{number of tallied trees}}{\text{number of sampling points}}\\ &= 10 \times \frac{240}{30}\\ &= 80 \text{ square feet}\end{aligned}$$

Pulpwood Estimates

Suppose, further, that the trees on the area are pine pulpwood size in two even-aged stands. Qualifying trees would be counted in the same manner as described earlier, but each age group should be tallied separately and the average total height of each group should be recorded.

For example, 100 tallied trees might have had an average total height of 50 feet, whereas 140 tallied trees might have had an average total height of 80 feet, assuming that the same 30 sampling points were used.

$$\begin{aligned}\text{Estimated standard cords per acre} &= \frac{(\text{number of tallied trees})(\text{average height in feet})}{(20)(\text{number of sampling points})}\\ &= \frac{(100)(50) + (140)(80)}{(20)(30)}\\ &= \frac{16{,}200}{600}\\ &= 27 \text{ standard cords per acre}\end{aligned}$$

Why This System Works

The gauge creates an optical angle of 104.18 minutes with its vertex or intersection at your eye. The 104.18-minute angle gauge tallies all trees closer to the sampling point than 33 times tree dbh. Whatever the size of the tallied tree, the instrument guarantees that it is within a plot radius determined by that particular tree's dbh. Thus, a tree with a dbh of ½ foot (and a basal area of $\frac{1}{5}$ square foot) would be tallied inside a plot with a radius of 16½ feet (and an area of $\frac{1}{50}$ acre). The basal area of each 6-inch tree, then, must be multiplied by 50 to place it on a per-acre basis. This means that such a tree would contribute $(50)(\frac{1}{5}) = 10$ square feet to the estimate of basal area per acre. However, a tree with a dbh of 1 foot (and a basal area of $\frac{4}{5}$ square foot) would be tallied inside a plot with a radius of 33 feet (and an area of $\frac{4}{50}$ acre). This means that each 12-inch tree would contribute $(\frac{50}{4})(\frac{4}{5}) = 10$ square feet to the estimate of basal area per acre.

Tallied trees of different sizes would have different basal areas, different plot areas, and hence different blowup factors, but the product of each tallied tree's basal area times its appropriate blowup factor would always be 10 square feet—the constant contribution of any tallied tree to the estimate of basal area per acre. This is why the diameter of a tallied tree and its distance from the sampling point are immaterial.

Chapter 7

FIRE AND THE FOREST—*IS IT HOT IN HERE?*

This unit adapted from *A Guide for Prescribed Fire in Southern Forests*.

OBJECTIVES

After completing this lab exercise, students should be able to:

- explain the conditions needed to conduct a prescribed burn
- complete a plan for a controlled burn

BACKGROUND

A fire in the forest can be a catastrophe, or it can be a great benefit. Burning is sometimes used in managing a forest to control understory and debris that might accumulate and cause a catastrophic fire. **Prescribed burns** have to be carefully planned to avoid damage to the trees and the surrounding environment. Certain weather conditions must be met, and the conditions in the forest must be right. In Exercise 1, you will be able to test the condition of a forest and make plans to do a **controlled burn**.

In preparing for and carrying out a prescribed burn, several factors must be taken into consideration. One of the most important is that of weather. A dry or windy condition can cause problems in controlling a prescribed burn. The information for planning a prescribed burn can usually be easily obtained. Local National Weather Service offices (the National Oceanic and Atmospheric Administration, or NOAA) will furnish weather forecasts and outlooks. In addition, some states have the services of fire-weather forecasters, who can give spot forecasts for the area to be burned. Current forecasts and outlooks are also available at any office of the state fire control agency. These sources can supply fire danger data from a source reasonably near the area to be treated.

Field observations of weather should be made at or near the prescribed burn area before and during burning. Such observations serve as a check on the applicability of the forecast and keep the burning crew up to date on any changes or effects of local influences. Handy belt weather kits, which contain a **psychrometer** and wind-speed measuring instrument, are useful. With this kit, and by observing cloud conditions and other weather indicators, a competent observer can obtain a fairly complete picture of current weather.

Wind

Prescribed fires behave in a more predictable manner when some wind *movement* is present. Ideal conditions occur when both wind speed and direction are steady. On-site winds vary with the density of the stand and with crown height. The preferred range for wind speeds in the stand is 2–10 mph for most fuel and topographic situations. These are the most desirable winds for prescribed burning, but specific conditions may tolerate higher speeds. High wind speed will dissipate the heat of a backfire quickly, resulting in less damage to the overstory than a fire backing into a low-speed wind. Wind speeds should be in the low to midrange when head fires are used. This way, high fire intensities caused by extreme rates of spread are prevented, yet enough wind is present to keep the heat from rising directly into tree crowns. Open areas with no overstory can be burned if there is little or no wind.

Perhaps of greater importance than wind speed is the length of time the wind blows from one direction. Persistent wind directions occur frequently during winter in high-pressure areas after the passage of a cold front, when winds are typically from the west or northwest. Winds from the southwest may also be dependable. East winds are not reliable for prescribed burning in many parts of the South except for the coastal areas.

The most critical areas, with regard to fuel and topography, should be burned when persistent wind directions occur. The relatively easy burns may be attempted under less desirable wind conditions. Sea breezes, topography, and local effects may have a bearing on favorable wind conditions, so these factors should be considered in planning the burn. For example, in areas with high or rolling hills, topography, not the wind, may be the chief factor controlling the direction and rate of fire spread.

Relative Humidity

Relative humidity is an expression of the actual amount of moisture in the air compared to the total amount the air is capable of holding at that temperature and pressure. The amount of moisture in the air can have a direct effect on the rate of burn for fuels in the prescribed burn area. A temperature rise of 20°, which might occur from sunrise to midafternoon, reduces the relative humidity by about half, and a temperature drop in early evening causes relative humidity to rise. When a cold front passes over an area, the temperature drop is usually accompanied by a drop in humidity. The lower humidity is the result of a change in air mass from warm and moist to cold and dry.

Preferred relative humidity for prescribed burning varies from 30–50 percent. Under special conditions, a wider range of relative humidity (as low as 20 percent and as high as 60 percent) can produce successful burns. When relative humidity is as low as 20 percent, prescribed burning is dangerous because fires are more intense and spotting is more likely. When the relative humidity is 60 percent or higher, fires may not burn an area completely or may not burn hot enough to accomplish the desired result.

The moisture content of fine fuel responds rather rapidly to changes in relative humidity (see Figure 7–1). However, because there is a time lag involved for fuels to achieve equilibrium with the moisture condition of the surrounding atmosphere and because previous drying or wetting regimes can influence fuel moisture, relative humidity by itself cannot be relied on completely for predicting the success or failure of a burn.

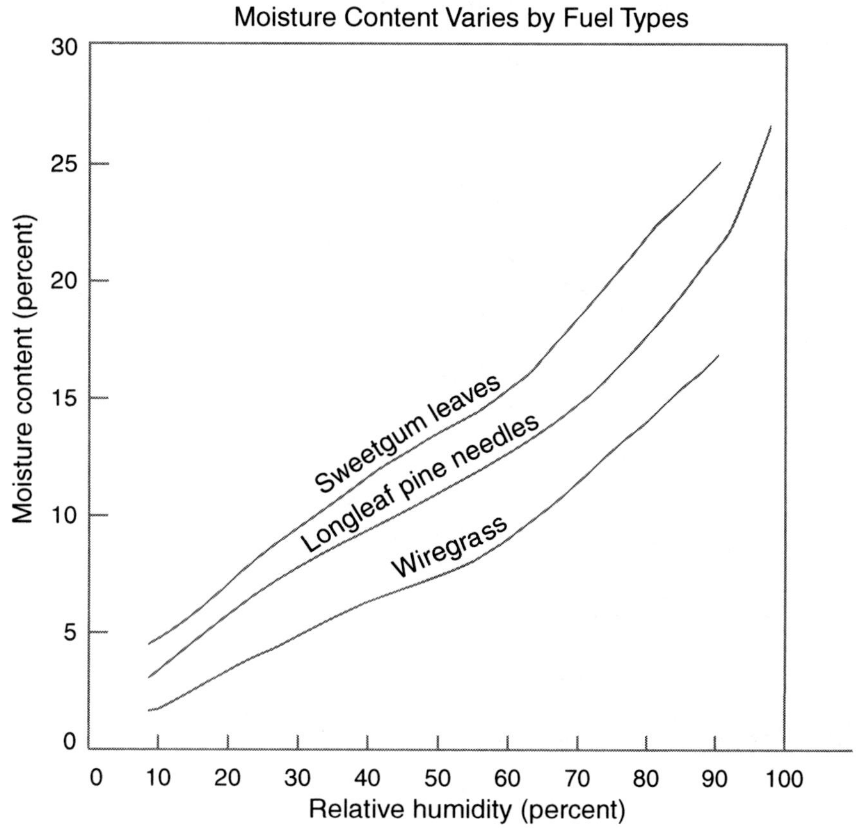

Figure 7–1 *(Courtesy United States Forest Service)*

Temperature

Air temperature can also have an influence on the rate of burn and the temperature reached by the fire. Low air temperatures of 20°–50°F are recommended for winter burning because there is less chance of raising foliage or stem tissue to lethal temperatures. Because killing temperatures are easier to reach, air temperatures of 80°–95°F are recommended when summer burning is used to control undesirable species. The average lethal temperature for cambium or unprotected tissue is 135°F.

The effect of temperature on moisture changes in forest fuels is important. When sunshine falls on the fuels, there are wide differences between the air and fuel temperatures. Moisture moves readily from the warmer fuel to air, even though the relative humidity of the air may be high.

Rainfall

From ½–1 inch of rain is desired 1–3 days before a prescribed burn. This usually ensures that the soil will be moist and deep organic layers will not be burned. Damp soil also protects small tree roots and microorganisms. This is a general rule; certain situations may call for dry duff if the burn is to accomplish a specific objective such as exposing mineral soil for seedbed preparation or eliminating debris in site preparation.

Rainfall affects fuel moisture and soil moisture (see Figure 7–2). In winter, rainfall may be fairly easy to forecast throughout the South; but in summer, when shower activity prevails, the prediction of rainfall at individual locations becomes very difficult. For this reason, the prescribed burner needs some idea of amounts of rain falling on the area to be burned.

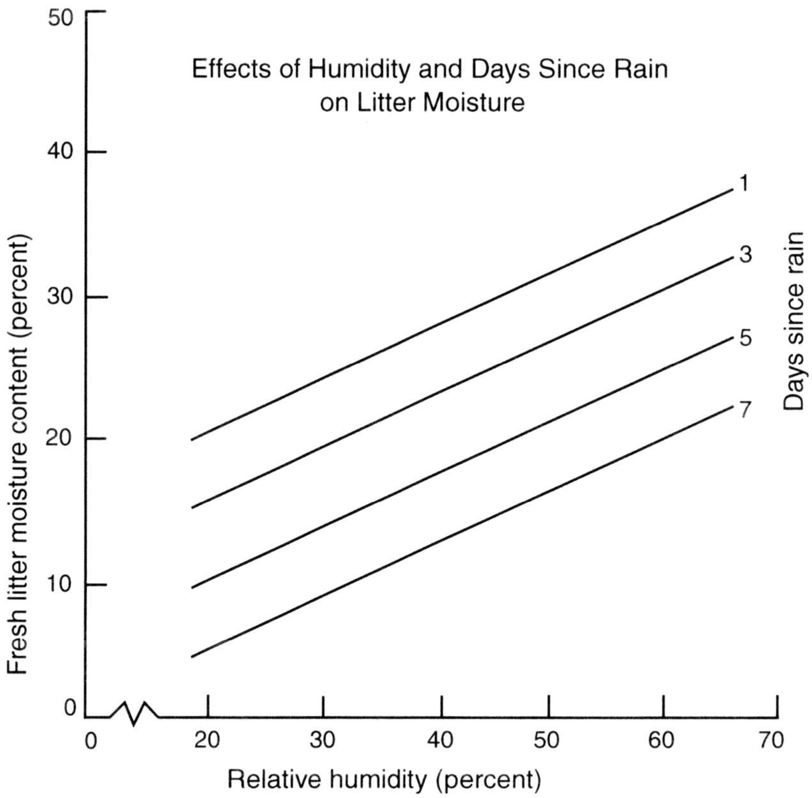

Figure 7–2 *(Courtesy United States Forest Service)*

Firing Techniques

Various firing techniques are available to help accomplish a burn objective under different weather, topography, and fuel conditions. Burning objectives, fuels, and weather factors must be correlated closely with the proper firing technique to prevent harmful effects to the forest resources. Atmospheric conditions should be favorable for dissipating smoke into the upper air levels and away from smoke-sensitive areas such as highways, airports, and urban areas.

Based on behavior and spread, fires either move in the same direction as wind (**head fire**), in the opposite direction to the wind (**backfire**), or at a right angle to the wind (**flank fire**). Any fire can be described by these terms; for example, a spot fire would develop all three types if allowed to burn long enough. The head fire is the most intense because of its faster spread rate, wider burning zone, and greater flame lengths. The backfire is the least intense, having slow spread rates, a narrow burning zone, and short flames. The flank fire's intensity is intermediate.

When slight variations in fuel volumes or prescribed weather conditions occur, consider combining two or more firing techniques to achieve the same result.

Backfire

The backfire technique consists primarily of backing fire into the wind. Fire is started along a prepared baseline—such as a road, plowed line, stream, or other form of barrier—and allowed to back into the wind. Research indicates that variations in wind speed have little effect on the rate of spread of fire burning into the wind. Normally, such fire proceeds backward at a speed of from 1–3 chains per hour. This technique is perhaps the easiest and safest type of prescribed fire, provided a steady wind (speed and direction) prevails. It produces minimum scorch and lends itself to use in heavy fuels. Some disadvantages are the time-consuming progress of the fire and the need for interior-line plowing at frequent intervals, usually every 10–20 chains, to ensure burning out of the strips. Stronger wind velocities (4–10 mph, ground level) are desirable in using this firing technique so that smoke is well dissipated and heat is prevented from rising directly into tree crowns. Burning on slopes has an effect similar to wind on rate of spread. Downslope fires burning in hilly areas, for example, would give much the same results as backfires in flat country.

Strip-Head Fire

The **strip-head fire** involves setting a line or series of lines of fire upwind of a firebreak in such a manner that no individual line of fire can develop to a high energy level before it hits either a firebreak or another line of fire. The strips vary in width, depending on stand density; type, amount, and distribution of fuel; and desired results. Usually, strips should be about 1–3 chains apart.

Frequently a combination of strip-head fire and backfire is used to treat an area (see Figure 7–3). The base control line would be treated with a backfire and then the remainder of the area treated with strip-head

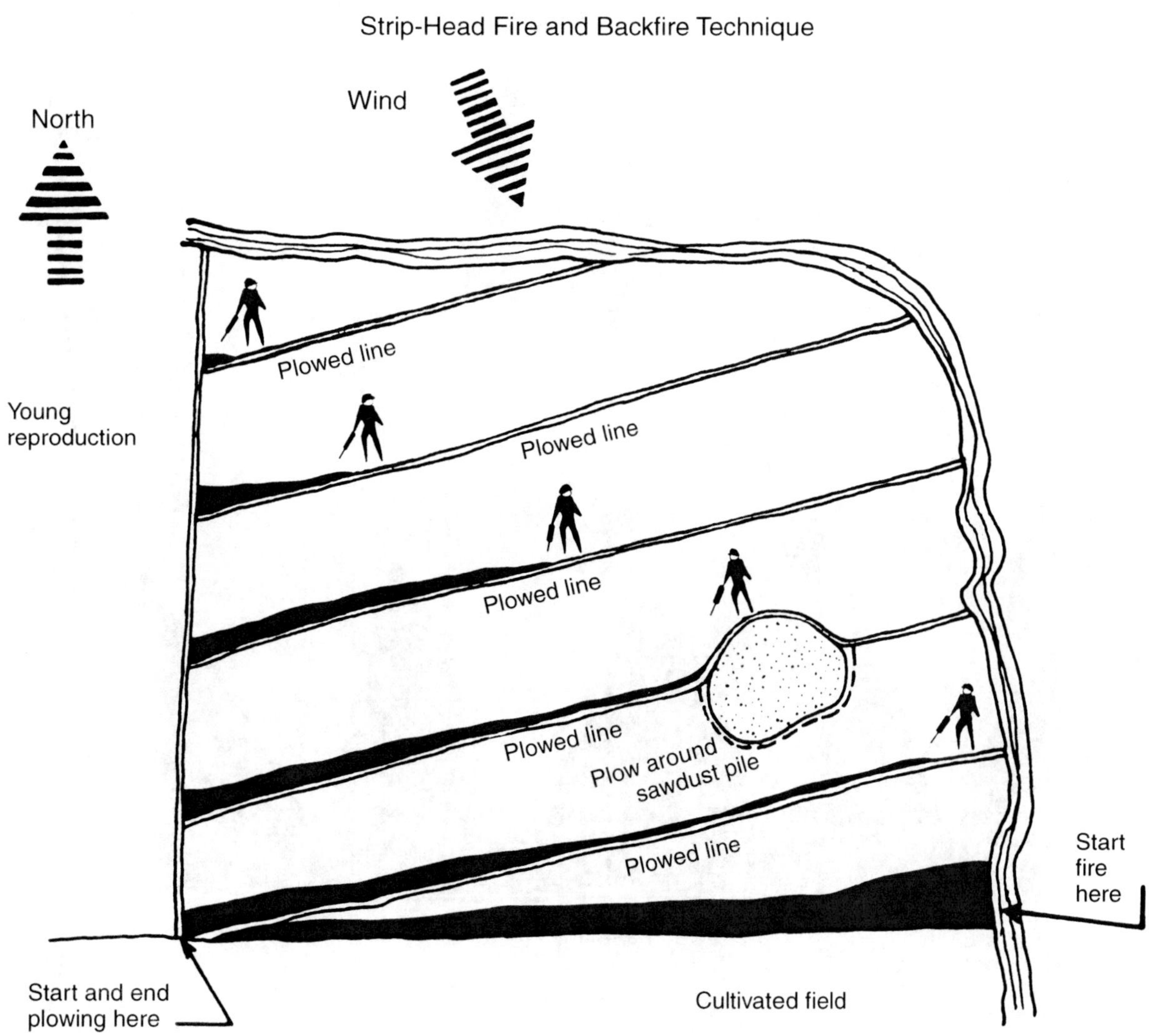

Figure 7–3 *(Courtesy United States Forest Service)*

fires. Compensation for minor changes in wind direction can be made by altering the angle of strip fire with the baseline. Also, adjustments for fuel density and composition can be made by altering strip widths. This method permits quick firing and smoke dispersal under optimum conditions.

Occasionally, on small areas with light and even fuel distribution, a head fire may be allowed to move over the entire area without stripping to better accomplish the objective. Where this method is usable, it cuts down on the areas of increased fire intensity that would occur at the junctures of strip fires burning together.

Flank Fire

The flank-fire technique consists of treating an area with lines of fire set into the wind, which burn outward at right angles to the wind (see Figure 7–4). This technique also requires considerable knowledge of fire behavior, particularly if used by itself. It is used quite often to secure the flanks of a strip-head fire, backfire, or other methods as they progress. It is sometimes used to supplement a backfire in light fuel areas or under more humid weather conditions. This method of firing can stand little variation in wind direction and needs expert crew coordination and timing. It is useful on a small area or to facilitate burning a large area in a relatively short time.

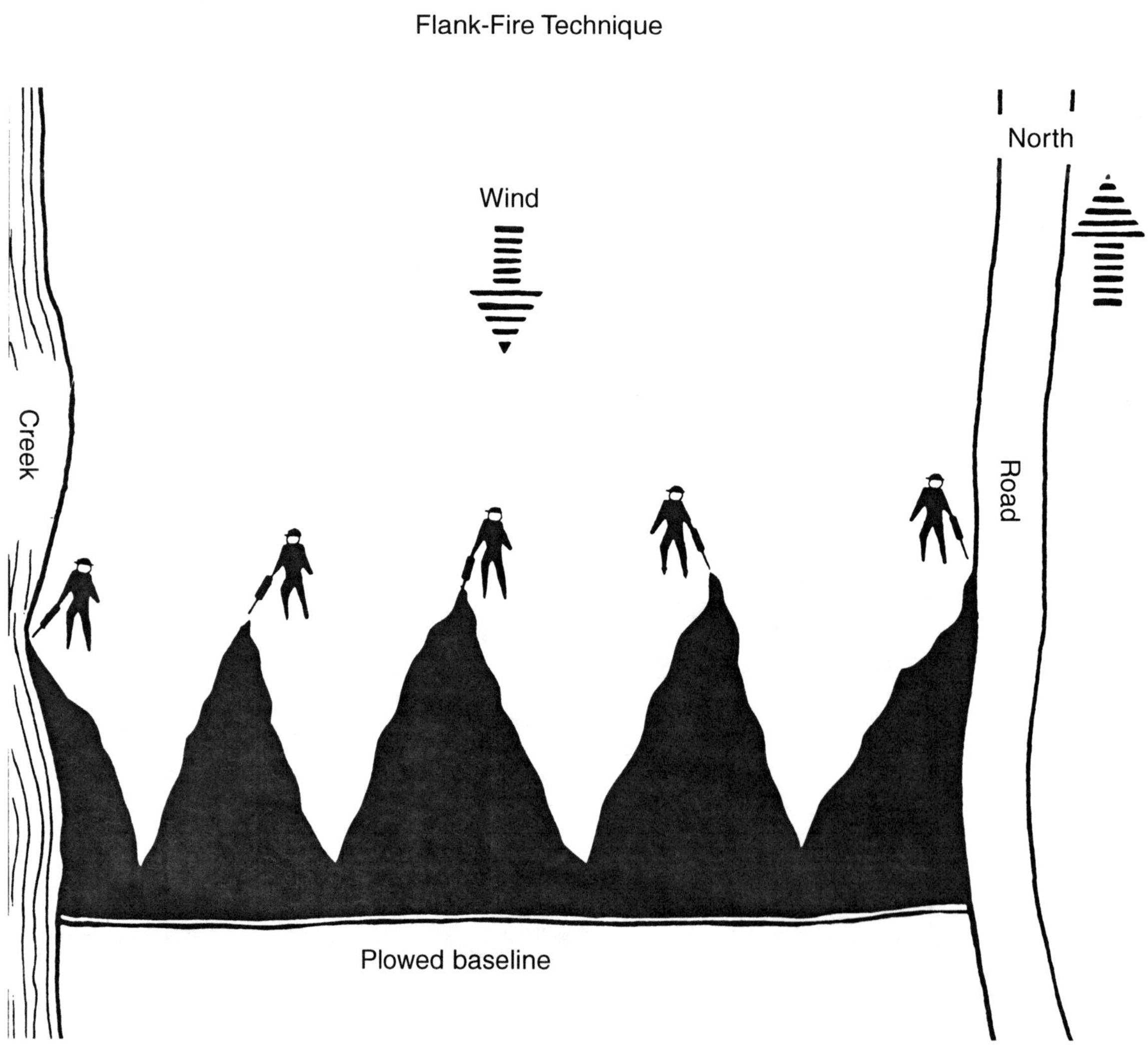

Figure 7–4 *(Courtesy United States Forest Service)*

Spot Fire

The technique of spot firing is an exact procedure that requires considerable experience by the prescribed burner. This method employs a series of small **spot fires** that burn in all directions as they come together, minimizing the possibility of any one spot gaining sufficient momentum to start a hot run. Timing and spacing of the individual fire spots are the keys to successful application of this method (see Figure 7–5). When this technique is used in a residual stand, care should be taken not to space spots closer than 2 or 3 chains. If the spots are closer, more junction zones of increased intensity are created. If the spots are too far apart, individual spots will build up hot fire heads. A skilled crew can treat a large area in a short time using this technique. Inexperienced prescribed burners should use some other technique until they gain considerable experience in fire behavior. This method can be used with light variable winds.

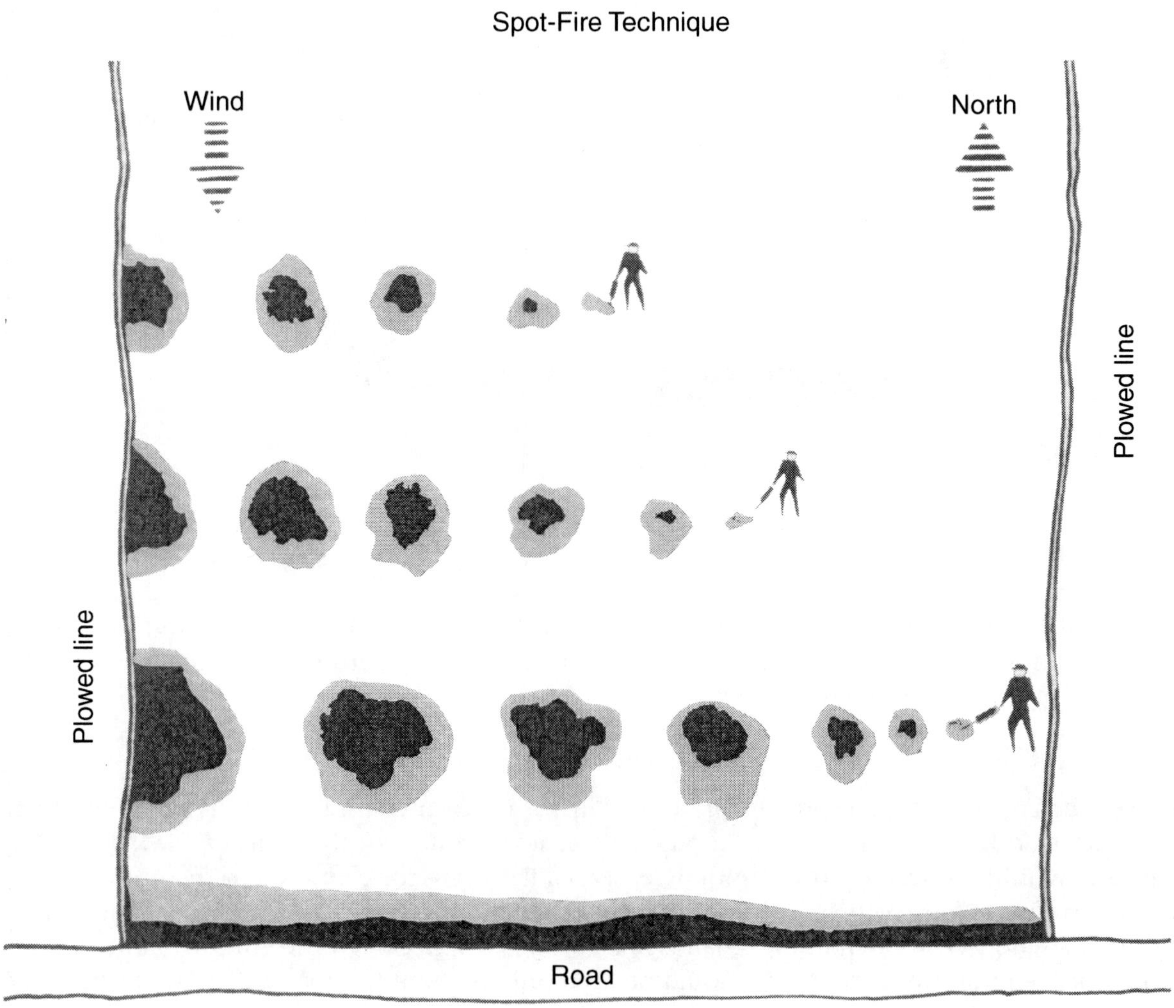

Figure 7–5 *(Courtesy United States Forest Service)*

Center and Circular (or Ring) Firing

Also known as "ring firing," **center and circular firing** is useful particularly in slash-disposal efforts on cut-over regeneration areas preparatory to seeding or planting. This method of burning is used on areas where a relatively hot fire is needed to reduce or eliminate logging slash and debris. As with other burning techniques, a base control line, usually on the downwind side of the burn area, is made secure with back and flank firings. Once the baseline is secured, the entire burn area is circled with fire and allowed to sweep over the area (see Figure 7–6). Often, during periods of light wind, one or more spot fires near the center of

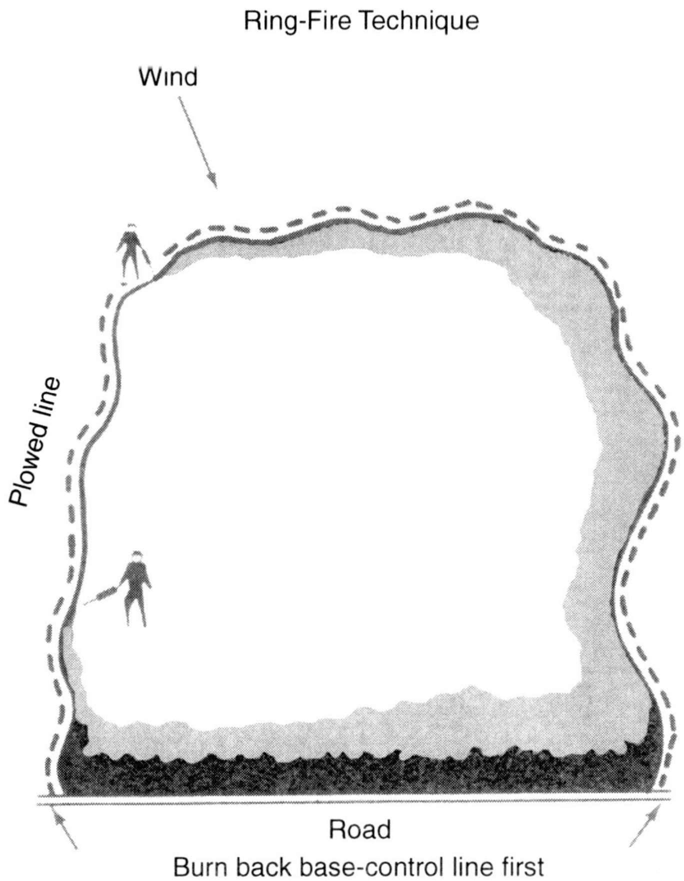

Figure 7–6 *(Courtesy United States Forest Service)*

the area can assist in pulling the outer circle of fire toward the center. This firing method can generally be used in any season, as weather conditions are not as critical. However, caution is in order, particularly when weather conditions border on the unstable.

Chevron Burn

The **Chevron burn**, another technique of prescribed firing, is designed for use in hilly or Piedmont areas to fire ridge points or ridge ends. Lines of fire should be started simultaneously from the apex of a ridge point and progress downhill. Basically this technique involves a flank-fire concept, except the lines of fire are not parallel. Any two lines of fire converge at a point, rather than along a line. This results in the junction zones or areas of increased intensity remaining small (see Figure 7–7). The Chevron technique alone is not suited to most hilly or Piedmont topography; it should be used only in combination with other techniques.

Smoke Management

Prescribed burning, however useful it may be in accomplishing certain resource objectives, can contribute to air pollution. The use of prescribed burning carries with it an obligation to eliminate or minimize any adverse environmental effects. Listed below are some guidelines to observe in reducing the impact of smoke:

1. *Have defensible objectives.* Be sure you have clear resource objectives and have considered the impact on the total environment, both on and off site.
2. *Obtain and use weather forecasts.* Weather information and fire-weather forecasts are available to all resource managers through the state forestry agency. Be sure to use them. Such information is needed to determine what will happen to the smoke, as well as to determine the behavior of the fire.

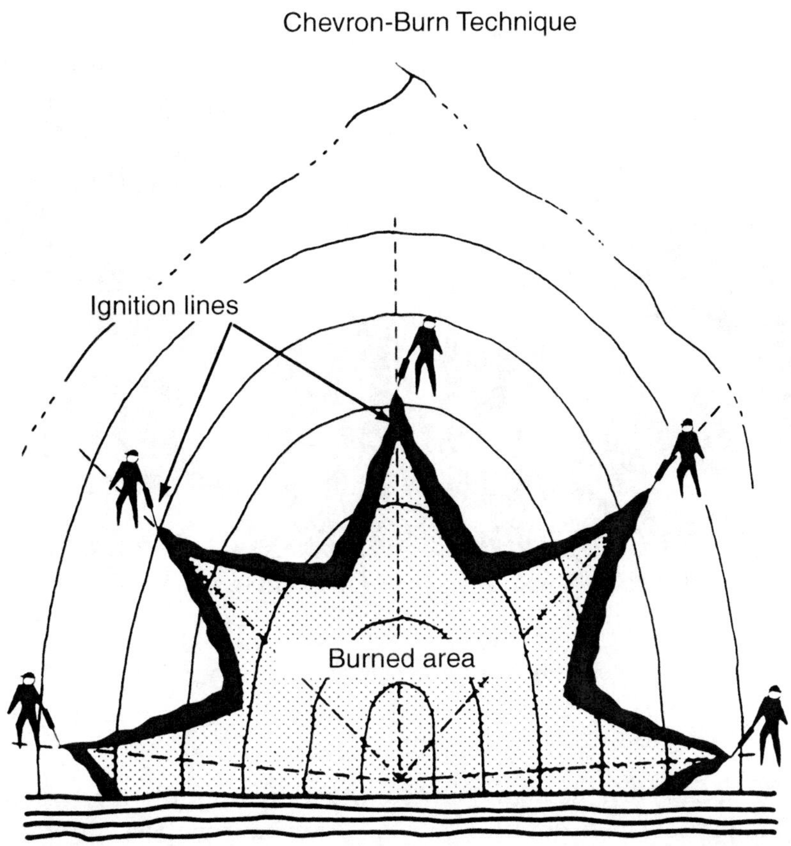

Figure 7–7 *(Courtesy United States Forest Service)*

3. *Do not burn during pollution alerts or temperature inversions.* Smoke will tend to stay near the ground and will not disperse readily. Many fire-weather forecasters include this in their regular forecasts.

4. *Comply with air pollution control regulations.* Know the regulations when you make the prescription. Check with the state fire control agency.

5. *Burn when conditions are good for rapid dispersion.* The atmosphere should be slightly unstable so smoke will rise and dissipate, but not so unstable as to cause a control problem (see Figure 7–8). Again, your fire-weather forecaster can help.

6. *Determine the direction and volume of smoke, especially near highways and populated areas.* Smoke management guidelines will help in making this determination.

7. *Use caution when near or upwind of smoke-sensitive areas.* Burning should be done when wind will carry smoke away from heavily traveled roads, airports, and populated areas.

8. *Notify the local fire control office, nearby residents, and adjacent landowners.* This is not only common courtesy, but it is also a requirement in most areas. Notification will let these people know it is not a wildfire, and you will get advance notice of any adverse public reaction—as from people with respiratory ailments, those with washdays, and so on.

9. *Use test fires to confirm smoke behavior.* Set them in the area proposed for burning, away from roads or other "edge" effects.

10. *Use backfires when possible.* Backfires give more complete consumption of fuel and produce less smoke. Even though they are slower and more expensive, backfires produce fewer pollutants and restrict visibility less than other firing techniques.

Smoke column under stable conditions

An unstable atmosphere

Figure 7–8 *(Courtesy United States Forest Service)*

11. *Burn in small blocks.* The larger the area being burned, the more visibility is reduced downwind and a higher concentration of particulate is put into the air. However, it may be better to burn all the area needed when weather conditions are ideal for rapid dispersion of smoke.
12. *Mop up along roads.* Burn out and start the mop up along roads as soon as possible to reduce impact on visibility.
13. *Be cautious of nighttime burning.* Predicting smoke drift and visibility is more difficult at night. The wind may lessen or die out completely, and smoke will then tend to stay near the ground. Burn at night only when you have a definite forecast of optimum conditions.
14. *Have an emergency plan.* Be prepared to control traffic on nearby roads if wind changes direction. Be prepared to plow out and stop a prescribed burn if it is not burning according to plan or if weather conditions change.

EXERCISE 1 BURN, BABY, BURN—DEVELOPING A PLAN FOR A PRESCRIBED BURN

Materials Needed

- Access to your area's regulations for conducting a controlled burn
- Access to a local weather station to obtain weather conditions
- A forest plot in need of a prescribed burn

Procedures

Contact the nearest office of your state's Forestry Commission and obtain a copy of its regulations and conditions necessary for conducting a prescribed burn. Locate a forest plot near your school that you can use to plan a prescribed burn. Call the weather station nearest you and obtain the weather conditions for the day you visit the site. Use the form provided in this manual (or a similar form obtained from your Forestry Commission) and fill in the data for your plot. Make a determination of when a prescribed burn should be done. Be sure to draw a map of the area in the space provided at the end of this chapter.

Step 1 *Planning the Prescribed Burn*

Prescription burning is a highly technical job requiring a knowledge of fire behavior, suppression techniques, and the environmental effects of fire. A written prescribed burning plan prepared by a knowledgeable person is needed for each area to be burned.

Draw up the plan before the burning season, then burn when the prescribed weather materializes. Some plans may be quite short and simple, whereas others will be complex. The unit can vary from a few to more than 1,000 acres, but the topography and amount and type of fuel in a unit should be similar. It could, however, be made up of a series of blocks in the same compartment or management unit as long as the same objectives apply and the fuel is similar.

Break large areas down into one-day burning blocks or smaller. Use existing barriers such as roads and creeks as much as possible, but make sure they are still effective barriers at the time of the burn.

Step 2 *The Written Plan*

Use of a prepared form with space for all the information needed is the best approach. The form will serve as a checklist so you can make sure you do not overlook some aspect or potential impact.

The written plan should include the following:

1. Reasons for burning and desired objectives
2. Description and map of area
   - Weather needed
   - Publicity needs
   - Alternative weather and firing pattern, if any
   - Season and time of day
   - Firing method
   - Control and mop-up
   - Appraisal of prescribed burn

Step 3 *Purpose and Objective*

Include in the written plan the purpose or reason for prescribing a treatment using fire; for example, brownspot control, hazard reduction, wildlife habitat, and so on. In addition, give a specific objective. State exactly what the fire is to do, what it should kill or consume, how much litter should be left, and so on. Also give a concise explanation of fire behavior expected, such as how high the flames should be, and how intense. This will enable the person in charge of the burning to possibly vary the technique and still accomplish your objectives if the prescribed weather conditions do not materialize. Such information will also be useful in determining the success of each burn. In addition, include alternative weather conditions (wind, relative humidity, and/or fuel moisture) with a method of burning that will produce approximately the same intensity fire and accomplish the desired objectives.

Step 4 *Fuel Conditions*

The amount of fuel and the intensity of burn desired will determine the weather factors needed and the burning technique to use. The species and height of the overstory will determine the maximum intensity that can be tolerated. Where large amounts of fuel are present, cooler burns can be accomplished by burning when the humidity and fuel moisture are in the higher range so a smaller percentage of the fuel will be available. Backfire will also result in a less intense fire. Lower air temperatures are desirable with a more intense fire, especially when the understory fuels are tall. Two separate prescribed burns may be necessary to eliminate heavy volumes of fuel without damage to the overstory.

Step 5 *Description of the Burning Unit*

Include the following descriptions in the plan:

1. Location of area and number of acres to burn
2. Record of previous burnings
3. Complete description of the area, including
   - *Overstory:* Type, density, and size
   - *Understory:* Type, density, and height
   - *Fuels:* Type, density, age, and volume
   - *Soil type* and *topography*

Step 6 *Map of the Burning Unit*

A detailed map of each burning unit is an important part of the burning plan. The map can be used to show the unit boundaries, landownership, topography, areas to be plowed out, firing and holding details, and other essential information. Show areas that should be excluded or protected, such as improvements, young reproduction, sawdust piles, and so on. Subdivide each area to be burned into logical, one-day burning blocks or smaller blocks if smoke management needs so dictate (see Figure 7–9).

Step 7 *Weather Factors*

Weather factors that need to be considered are:

1. Wind (direction and speed)
2. Relative humidity
3. Fine fuel moisture
4. Temperature
5. Burning index range
6. Stability
7. Dampness of litter

Step 8 *Season and Time of Day*

Winter. Most prescribed burning is done during the winter months so the larger stems of the overstory will not be killed. In heavy growth, the area can be treated with a low-intensity backfire and followed with another the next winter.

Spring. More variable weather and generally higher fire danger dictate smaller burns for greater flexibility during spring. Check with the local wildlife specialists to determine the exact period when prescribed burning would be detrimental to wildlife.

Summer. Because the air temperature is generally high during the summer, it takes much less heat to raise the cambium to lethal temperature. For this reason, summer burns are used to kill undesirable hardwoods—usually a series of burns after a winter burn.

Time of Day. Normally, plan burning operations so the entire prescribed burning job can be completed within a standard workday. The last job of the day is to patrol and mop up the perimeters of the fire.

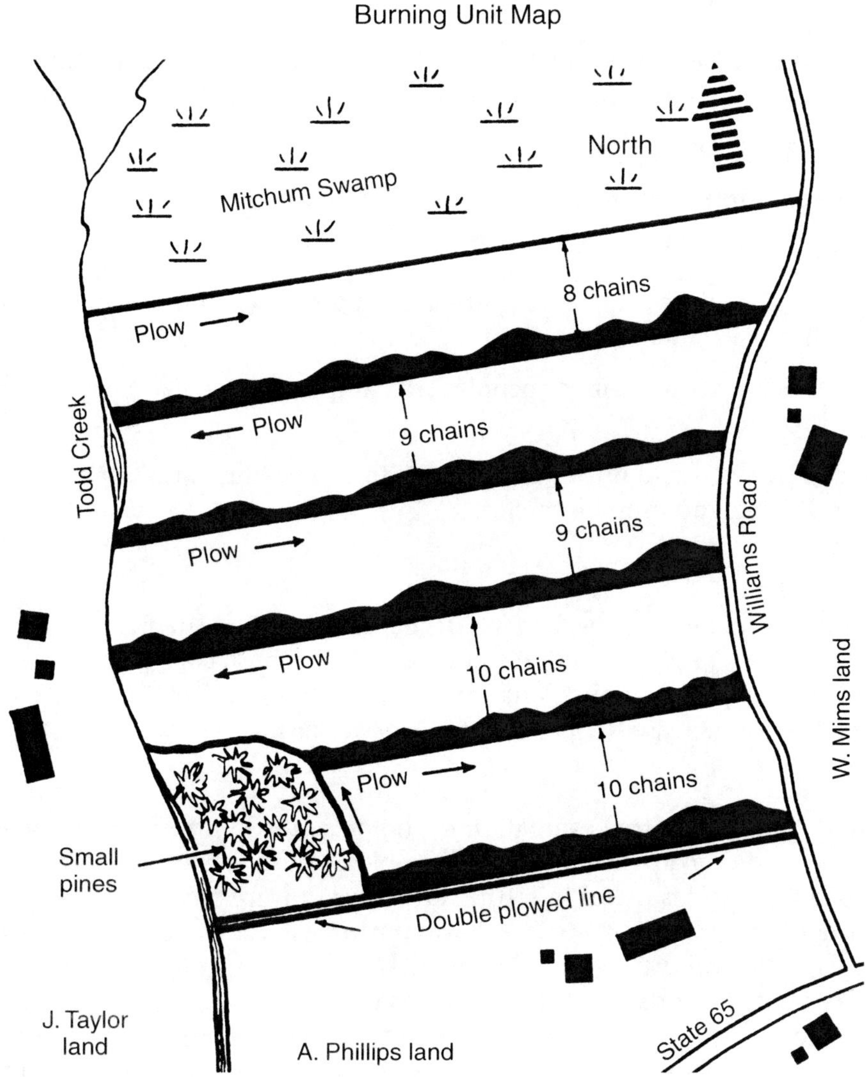

Figure 7–9 *(Courtesy United States Forest Service)*

Generally, burning during daylight hours is better because constant and steady winds occur then. At night, smoke tends to stay close to the ground and collects in small depressions. Also, the relative humidity usually goes up at night, resulting in spotty burning. However, some winter nights have strong cold fronts moving in when the wind persists and the relative humidity does not rise. When night burning is done, keep a close check on wind, humidity, and smoke drift.

Fall is generally considered to be too dry for prescribed burning. Most prescribed burns are conducted during the other seasons.

Step 9 *Public Considerations*

Publicity on Burning. List in the plan the name and telephone number of the local state fire protection officer and neighbors who should be contacted prior to the prescription burn.

Impact of Smoke. List in the plan any sensitive areas near or downwind of the area to be burned. Include precautions needed to lessen the impact from the smoke.

Legal Requirements. List any legal requirements that might apply and what the prescribed burner will need to do to comply with those requirements.

Step 10 *Firing Plan*

An important part of a successful prescribed burn is the plan for firing and holding a unit. The firing plan should consist of a narrative section, plus a detailed map. Include the following items as they apply:

1. Planned time of day for setting the fire
2. Firing method to use, firing sequence, and possible alternatives
3. Lines to plow and lines to fire
4. Human resources and equipment needed and planned distribution for setting, holding, patrolling, and mopping up the fire
5. Location and number of reinforcement people and equipment that can be mobilized rapidly if fire escapes
6. Instructions for overhead, including complete description or illustration of assignment and forces needed to fire out, hold, and mop up the fire

Step 11 *Control and Mop-up*

A well-prepared plan will include the necessary safeguards to confine the fire to the prescribed area and to reduce smoke impact. Mop-up will be prompt and complete. Emphasize protection of all adjacent lands. Consider and make backup plans for any variation in forecasted weather that may change a prescribed fire into a damaging wildfire or increase the pollution in smoke-sensitive areas.

Step 12 *Appraisal*

Include space in the plan for a written evaluation of the prescribed burn. A record of actual weather conditions, behavior of fire, and total environmental effects of the burn is essential in determining the effectiveness of the prescribed burn and is useful in setting criteria for future burns.

At the beginning of the prescribed burn and also during early afternoon, record wind speed and direction, fuel moisture, humidity, burning index, temperature, days since last rain, and dampness of soil and lower litter. Also record behavior data of the fire such as type of fire used, height of flames, and forward rate of spread. After the burn, record amount of scorch and bark char; consumption of brush, litter, and duff; and any other evidence of fire intensity such as cracks in bark, exposed mineral soil, and unburned areas. In addition, include a short narrative on the success of the burn.

Understory Prescribed Burning Unit:

Prepared by ______________________________ Permit no. ______________

Signature ______________________________ Date ______________

State __________ County __________ District __________ Company ______________

Gross acres __________ Net acres __________

Landowner ______________________________

Address & phone no. ______________________________

Person responsible & how to contact day and night ______________________________

____

(Draw map on the space provided below)

Record of Previous Burning: Date __________ Fire type __________ Results __________

Description of Stand:

Overstory: Type, density, size ______________ Height to bottom of crown ______________

Understory: Type, density, height ______________________________

Dead fuels: Type, density, age, volume ______________________________

Soil type and topography ______________________________

Purpose(s) of Burn: ______________________________

Specific Objectives: ______________________________

____

Preburn Factors:

Chains to plow (see map): Exterior __________ Interior __________ Total __________

Chains to fire (see map): Exterior __________ Interior __________ Total __________

Crew size __________ Equipment needs ______________________________

Estimated tons/acre __________ Total tons to be burned ______________

Ignition procedure (see map) ______________________________

Notify ______________________________

____

Regulations that apply ______________________________

Smoke-sensitive areas & critical targets (see map) ______________________________

____

Weather Factors:	**Desired Range**	**Predicted**	**Actual**
Surface winds (speed & direction)	______	______	______
Transport winds (speed & direction)	______	______	______
Stability/stagnation index	______	______	______
Minimum mixing height	______	______	______
Dispersion index	______	______	______
Minimum relative humidity	______	______	______
Maximum temperature	______	______	______
Fine-fuel moisture (%)	______	______	______
Days since rain ______ Amount	______	______	______
Burning index ______ Drought index	______	______	______

Fire Behavior:	**Desired Range**	**Actual Range**
Type fire	______	______
Best month to burn	______	Date burned ______
Time of day to start	______	Time set ______
No. of hours to complete	______	Completed ______
Flame length	______	______
Rate of spread	______	______
Fireline intensity	______	______
Inches of litter to leave	______	______

Evaluation Immediately after Burn:

Acres burned ______

Spotting ______ Distance ______

Any escapes ______

Objective met ______

Smoke problems ______

% of area with crown discoloration ______

5–25% ______ 26–50% ______ 51–75% ______ +76% ______

Live crown consumption ______

% understory vegetation consumed ______

Adverse publicity ______

Future Evaluations:

Evaluation by ______

Date made ______

Insects/disease damage ______

Date ______

Insect/disease damage ______

Crop tree mortality ______

% understory kill ______

Soil movement ______

Other adverse effects ______

______

Remarks ______

______

______

Map of Area:

In the space below, draw a detailed map of the area. Include property boundaries, roads, trails, streams, gullies, hills, and any characteristics that might affect the course of a fire or the plowing of a firebreak.

Chapter 8

WILDLIFE AND THE FOREST—*WHERE THE WILD THINGS ARE*

OBJECTIVES

After completing these lab exercises, students should be able to:

- explain the basic needs of all wildlife species
- understand how wildlife populations fluctuate over time
- explain how predator and prey populations are dependent on each other
- determine why environments can support only a limited number of individuals
- determine the wildlife population in a small area
- estimate the wildlife population in a larger area of similar type

EXERCISE 1 I'VE REACHED MY LIMIT—LIMITING FACTOR RELATIONSHIPS

Background

You have probably heard about the "balance of nature." The reality is that nature is never really in balance. Nature is always in flux as it reacts to one influence or the other. There are many conditions that can cause any one population of wildlife to increase and others to decrease, but they are never in balance; they are always changing. The most basic needs of wildlife include food, water, shelter, and space. A wildlife population also consists of predators and prey, the relationships between them, and the ecosystem they live in. This exercise will help you understand how wildlife populations fluctuate in a normal rhythm, as opposed to being in balance.

Materials Needed

- Paper
- Pencil
- Group of students
- Chalk or flagging tape to define a line

Procedures

To show students the dynamics of wildlife population change, the class will do a little role-playing. Here is how it works. The students will count off by 5. The 1s will be the predators; the 2s, food; the 3s, water; the 4s, shelter; and the 5s, space. Line the predators (group 1) up on one line. Next, line the other groups up in random order on the other line about 10 yards away. Members of the food group will identify themselves by putting their hands on their stomachs. The water group will identify themselves by holding their hands over their mouths. The shelter group will identify themselves by holding their hands on top of their heads. The space group will identify themselves by holding their hands behind their backs.

Students should understand **that in order to collect accurate data, they must follow the rules and keep track of their changing roles in the game**. Students should stand with their backs to each other and their hands in position. When the leader says "go," they are to turn around and wait for a predator to claim them.

Students in the predator line are to stand on the line and decide what they need to survive. They put their hands in the position to represent that need; and when the instructor says "go," they are to quickly walk across the space between the lines and claim one person who has his or her hands in the same position, thus representing the need they must fill. Predators cannot change their need until the next turn, even if that need is not available to them. Some of the predators will not find what they need, and they will become part of the ecosystem. Some predators will find what they need, and the students who return to the predator line now become predators.

The leader needs to count the number of predators and the number of students in the ecosystem after each cycle. The numbers will fluctuate as the game is played for 10–15 cycles.

Hypothesis

Write a statement predicting what you think will happen to the prey population when the predator population is high and vice versa.

____

____

Observations

Have a class discussion about the population changes that occurred from year to year (one year is equal to one cycle of the activity). Using the data from that discussion, try to determine why the populations of each group changed. The reasons for the changes are referred to as **limiting factors**.

Data

	20												
P	19												
R	18												
E	17												
D	16												
A	15												
T	14												
O	13												
R	12												
	11												
P	10												
O	9												
P	8												
U	7												
L	6												
A	5												
T	4												
I	3												
O	2												
N	1												
		Year 1	**Year 2**	**Year 3**	**Year 4**	**Year 5**	**Year 6**	**Year 7**	**Year 8**	**Year 9**	**Year 10**	**Year 11**	**Year 12**

Conclusions

1. Students will take this data and create a population graph representing the 10–15 years (cycles) represented. The vertical axis of the graph will represent the number of predators present in the ecosystem each year, and the horizontal axis will represent the number of years (cycles) that data were collected.

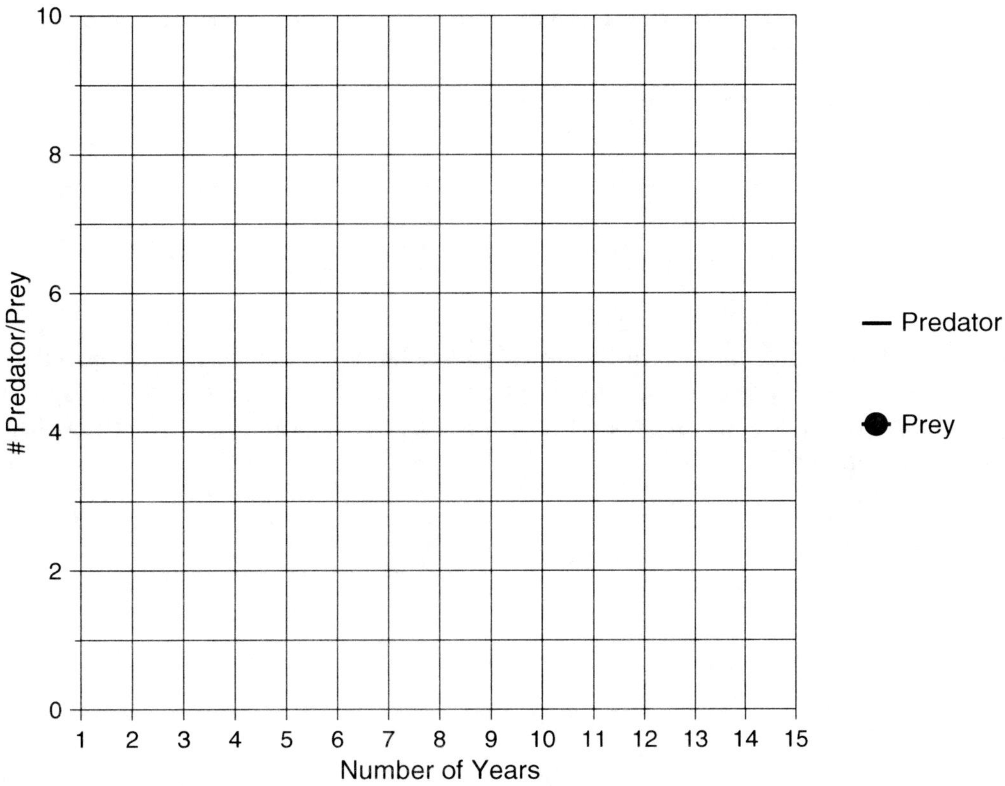

Figure 8–1

2. Explain why the population changes occurred the way they did from year to year.

3. How did limiting factors affect population declines and/or increases?

4. Look back to your hypothesis. With the data you collected, are you able to accept or reject your original hypothesis? Explain.

____

____

____

EXERCISE 2 THIS IS WILD—WILDLIFE POPULATION SURVEY

Background

Wildlife comes in all shapes and sizes, from the small to the large. To determine the population of wildlife in a given environment, the area is often studied in grids. The grid can be used as a representative sample to estimate the population in a given larger area. This exercise will help you determine the wildlife population of areas both large and small.

Materials Needed

- String
- Small stakes or nails
- Paper
- Pencil

Procedures

Step 1

Lay out a series of grid lines to create a square 3 × 3 feet (1 square yard). Define the squares by placing the nails or stakes in each corner, and then lay out string from corner to corner.

Step 2

Carefully search the grid and list each living animal that is found. Use the chart on page 71 to record your data. You may find several types of insects, perhaps mollusks like snails, as well as amphibians and reptiles. The entire group should visually survey the area around the grid locations. A list should be made of all the wildlife seen by the group. *Note:* Your presence may have scared off animals that normally live in that area, so look for signs like scat, scratches (areas were animals have clawed in the earth), and burrows. Count these signs as one animal. Be sure to keep in mind that larger animals like deer and elk are evaluated on a much larger scale; but for this lab, we will not take that into consideration. Also, the exact name is not important; so, writing something like "small black bug" is fine.

Once it totals the grids, the group can make an estimation of the population of a larger area by multiplying the known populations by the area of a larger space. For example, if you calculate the population of 10 square-yard grids, you can then estimate the population of a 100 square-yard area by multiplying the known population by 10.

Hypothesis

Write a statement predicting what type of animal you will find the most of.

____

____

Observations

Student Name: ______________________ **Grid Identification:** ____________

Species Observed

Animal Species	Insect Species	Mollusk Species	Reptile Species	Amphibian Species	Misc. Species

Conclusions

Look back at your hypothesis and the data you collected. Will you reject or accept your hypothesis? Explain.

____

____

____

EXERCISE 3 IT'S TOO HEAVY—CARRYING CAPACITY

Background

A given habitat has the ability to support a certain number of animals, but that number is limited by the available food, water, and shelter. This is called the **carrying capacity**, which can fluctuate from year to year depending on climatic conditions, wildfire, and other variables.

Procedures

Use different colored note cards to represent food, water, and shelter. Have each card represent 5 percent of the required amount of each limiting factor (food, water, and shelter). Each student will represent a unit of wildlife. This could be a deer, a bear, a cougar, a rabbit, or other creature. Distribute enough cards around a room or outdoor area so there is not enough of each limiting factor for each animal unit to survive.

Some animals may be injured or weakened from lack of food, water, or adequate shelter. One student will be assigned to collect cards with one hand behind his or her back. Another student will be assigned to hop from place to place while collecting cards to represent weak or injured animals.

After all the cards have been collected, each student should count his or her cards and see what percentage of each limiting factor has been accumulated. Students can use the chart on page 68 to record this information. The class should determine who survived and who did not (at least 1 casualty each round). The students who do not survive became part of the ecosystem and are out of the game. Those who did not survive should distribute their cards in the "habitat" again. The surviving wildlife units should again gather their cards. The number of survivors should be determined again, based on the percentage of each limiting factor collected.

Hypothesis

Write a statement predicting which limiting factor will have the most limiting effect.

___

___

Observations

Wildlife Student Name	Food Total % per round	Water Total % per round	Shelter Total % per round

Wildlife Student Name	Food Total % per round	Water Total % per round	Shelter Total % per round

Conclusions

After the exercise, have a class should discussion about why some wildlife survived and why some did not.

1. Compare the number of survivors during the first round to the number of survivors during the second round and explain the results.

___

___

___

2. Why was there a difference between these rounds?

___

___

___

3. What was the impact on weak or injured animals when it came to collecting food and to finding water and adequate shelter?

___

___

___

4. How did competition, or the lack of it, affect the number of animal units the habitat could support?

___

___

___

5. Look back at your hypothesis. Will you accept or reject it? Explain.

___

___

___

Chapter 9

WATER QUALITY AND WATERSHED MANAGEMENT— *KEEP YOUR HEAD ABOVE IT*

OBJECTIVES

After completing these lab exercises, students should be able to:

- determine what happens to surface water runoff around the school and the community
- explain the importance of surface water management to the ecosystem
- determine what impacts humans are effecting on local waterways
- explain what happens to water quality as streams pass through each community downstream
- take measurements in an aquatic environment and monitor changes that occur over time

EXERCISE 1 WHEN IT RAINS IT POURS—SURFACE WATER RUNOFF

Background

When it rains in most urban areas, surface water runs off roofs, paved roads, concrete surfaces, and other impervious surfaces. This water has to go somewhere. Take a look around your community and your school. Start with the roof over your head right now. When it rains, where does that water go? Look for rain gutters and downspouts. Although gutters and downspouts may direct water into the ground, that water must go somewhere.

Hypothesis

Write a brief statement predicting where you think the urban runoff ends up from your town or city.

____

____

Materials Needed

- Large graph paper or butcher paper
- Compass
- Pencil
- Tag board for making stencils
- Spray paint

Procedures

Step 1

Follow the water and see where it goes. Look into catch basins and see which way the pipes run. Check out ditches along roads and see where the water comes from and where it goes. If something washes down the storm drain, where does it end up?

Step 2

Use a large piece of paper to create a map. Place a directional indicator on the map. North should be at the top of the map. Draw a map of the water flow and graphically show where it comes from and where it goes. Use arrows to indicate the direction of water flow. Use dashed lines to indicate where the water is flowing underground.

Step 3

Determine how the elevation changes and explain how that influences the water flow on the map.

Step 4

List some of the pollutants that could logically be getting into the runoff.

1. ____
2. ____
3. ____
4. ____
5. ____
6. ____

7. ______________________________

8. ______________________________

9. ______________________________

Step 5

What are these contaminants most likely to pollute, and what are the impacts to wildlife and the environment?

____

____

____

Step 6

As a community or school service project, obtain permission to make stencils and paint warnings next to storm drains and catch basins. The warning should read "Dump No Waste, Drains to Stream" or "Drains to Wetland," or wherever your map showed the water ended up.

Conclusions

Did the run-off water end up were you hypothesized it would? Accept or reject your hypothesis.

____

____

EXERCISE 2 UP A CREEK—STREAM IMPACT CHECKOFF

Background

"We all live downstream" is a popular saying in talking about water quality. Anything that goes into the stream in one place flows downstream and impacts those environments and communities. It is important to analyze what factors are impacting waterways in a given area so those potential impacts may be recognized and corrected if possible.

Hypothesis

Before visiting a riparian zone (RZ) or beginning your online research, write a statement predicting whether urban, rural (farming), industrial, or other activities will cause the most harm to a RZ.

____

____

Procedures

Visit a stream and observe the riparian zone. Check off the existence of the following activities and determine if these activities impact the **riparian zone** (**RZ**), the area between waterways and dry land. If leaving the class room is not an option, find images on the Internet where RZs have been impacted by the activities listed below. Rate the amount of impact or damage, *0* for "no impact" up to *3* for "large impact."

0 No impact

1 Little impact

2 Moderate impact

3 Large impact

Urban Impacts

___ Road built in or very close to the RZ

___ Single-family home(s) encroaching into the RZ

___ Apartments, hotels or large nonresidential building close to the RZ

___ Construction of any kind that is near the RZ

___ Retail stores close to the RZ (think about potential runoff)

Rural (Farming) Impacts

___ Livestock impact from possible trampling and fecal contamination

___ Manure runoff from large livestock feedlots and dairies

___ Irrigation runoff that may include herbicides and pesticides

___ Cultivated soil that may easily be carried into the RZ

Industrial Impacts

___ Logging damage caused by large equipment, nonharvested felled trees, or overharvesting the RZ

___ Mining activities damage caused by sediment piles, chemicals, and equipment

___ Manufacturing activities damaging the RZ

Other Impacts

___ Forest run-off water causing pH changes in water from pine and fir forests

___ List and describe any other factors that can have a negative impact on the RZ:

____

____

Conclusions

1. Summarize the impacts that have been noted and write a statement explaining the long-term effects and what should be done about them.

____

____

____

____

____

____

____

2. What is happening to the water quality as the stream or river passes through communities downstream? Are they adding to the problem? Explain your answer.

3. Based on what you have learned, are you able to accept your hypothesis? Explain.

EXERCISE 3 DILUTION LAB

Background

As technology increases, we are able to measure potential pollutants in smaller and smaller concentrations. Many contaminants are now measured in parts per million, whereas they were previously measured in parts per thousand. Some chemicals are considered harmful in very small quantities; so, it is important to determine the potential exposure that is occurring in a lake or river. Fish or other aquatic organisms can ingest these contaminants, which can have negative impacts on their populations and on those populations that feed on them. To put contaminant levels in perspective, one part per billion is about one drop in a competition-size swimming pool.

Materials Needed

- Four clear, 1-liter bottles or 1-liter beakers
- Food dye
- Pipette or eyedropper
- Clean stir stick
- A white sheet of plain paper

Hypothesis

Read the procedures for this lab and then predict if you think you will be able to see any color change in the last bottle.

Procedures

Add 999 mL of clean water into three bottles or beakers. Add 1,000 mL of clean water to the fourth bottle. This last bottle will be your control. Use it to compare the other bottles to. When you compare bottles, use the white paper as a backdrop.

Step 1

To bottle A, add 1 mL or 20 drops of food dye (pollutant). Mix well.

Dilution: 1 mL/1,000 mL = _____ part per ______

Show your math:

Can you see any color change? ____

Step 2

To bottle B, add 1 mL or 20 drops from bottle A. Mix well.

Dilution: 0.001 mL/1,000 mL = ____ part per ____

Show your math:

Can you see any color change? ____

Step 3

To bottle C, add 1 mL or 20 drops from bottle B. Mix well.

Dilution: 0.000001 mL/1,000 mL = ____ part per ____

Show your math:

Can you see any color change? ____

Conclusions

1. Many pollutants are measured in parts per billion. Could something in this small an amount be toxic? Research this question and report your findings:

____

____

____

____

____

2. Could we possibly even tell if something is present in this small a quantity? Research this question and report your findings:

3. Write a brief statement accepting or rejecting your hypothesis. Explain.

EXERCISE 4 THE GLASS IS HALF FULL—WATER MEASUREMENTS

Background

When working near a waterway, you should be familiar with some of its features. For example, it is important to know the waterway's depth. Looks can be deceiving; and what may seem to be a shallow stream with no current, may surprise the unfortunate person who falls in unaware. Extremely cold water can cause a human body to quickly shut down, leaving a person unable to breathe or move normally. Another feature of a waterway that can be dangerous is its **sediment load**, which, if heavy, makes it difficult to see dangers under the water's surface. By taking time to quickly survey any waterway near your work or study area, your work in the forest will be safer.

Materials Needed

- Ping-pong ball
- Stopwatch
- Thermometer
- Clear glass jar

Hypothesis

After reviewing the procedures below, write a hypothesis stating why changes may occur in a waterway.

Procedures

Step 1

Visually survey a stream, pond, or wetland area.

Step 2

Take the measurements requested in the chart below and note them in the chart.

Observations

Width of Waterway	Depth of Waterway	Water Temperature	Water Velocity	Sediment Load
Location 1 width:	Depth at near edge:	Temperature at surface:	Velocity near edge:	Sediment load near edge:
Location 2 width:	Depth at near center:	Temperature at bottom:	Velocity at midstream:	Sediment load midstream:
Location 3 width:	Depth at near far edge:	Temperature at mid-depth:	Velocity at far edge:	Sediment load far edge:
Average width:	**Average depth:**	**Average temp:**	**Average water velocity:**	**Average sediment load:**
Notes: To determine water velocity, record the time it takes to float a ping-pong ball from one end of a known area to another. Determine the number of yards per minute or meters per minute. You can then calculate miles per hour.				
To determine sediment load, consider small soil particles in clear water to be "low," small to medium soil particles and other organic matter in slightly cloudy water to be "medium," and heavily muddied water with large particles suspended in it to be "heavy."				

Step 3

If possible, keep monthly records to determine how these measurements change over the course of a year. Speculate as to why these changes occur. Explain your observations below or explain why they might change over time.

____

____

____

____

____

____

Conclusions

Based on your measurements and observations, could you accept or reject your hypothesis? Explain.

____

____

____

____

Chapter 10

THE ROLE OF GOVERNMENT IN FORESTRY—*LAISSEZ FAIRE?*

OBJECTIVES

After completing these lab exercises, students should be able to:

- identify controversies involved in forest management
- analyze arguments on both sides of an issue
- present oral and written discussion
- work out compromises

BACKGROUND

Since its beginning, the U.S. government has had a role in the management of forests. Government regulations have almost always been controversial, and not everyone can be pleased with the controls the government implements. As with most controversies, there are two sides to forestry issues. Most often, the real role of government is to find the central ground and foster compromises between groups with opposing views.

PROCEDURES

In the exercises that follow, scenarios depicting real controversies that require government intervention have been provided. The class should divide into two groups, each group choosing a side in each controversy. Research and discuss all the arguments either for or against the issue. In the space provided, list your arguments. Then choose a spokesperson for each group and present the arguments in the form of a debate. After the debate, the two groups should discuss possible compromises and reach a consensus for each issue.

EXERCISE 1 THIS JOB IS FOR THE BIRDS

In the hills of West Virginia, a logging company has just moved into a small town. The company has been awarded a very large contract to cut timber on a 5,000-acre tract of land. According to the terms of the contract, the timberland is to be clear-cut and reseeded. A sawmill will be erected in the town, and 86 people will be hired to run the mill. In addition, more than 50 other workers will temporarily move into town. All the new jobs and people spending money locally will be a very big boost to the town's economy.

Just before the logging is to begin, a wildlife biologist discovers that there is a group of very rare woodpeckers living in the forest, and she is afraid that clear-cutting the timber will result in the destruction of the woodpeckers' habitat, thus leading to the loss of an endangered species. The dilemma is whether to continue with the cutting and risk losing an endangered species or cancel the operation and lose all the jobs and the boost to the economy that would come with the operation. One side should give reasons and arguments for continuing with the operation, and the other side should provide reasons and arguments for canceling the operation to save the woodpeckers. Think through each argument and list them below.

Reason or Argument 1

____

____

____

____

____

____

____

Reason or Argument 2

____

____

____

____

Reason or Argument 3

Reason or Argument 4

Reason or Argument 5

EXERCISE 2 DIG, BABY, DIG

Geologists in southern Utah have discovered that there is a vast supply of crude oil about a mile below the ground. The supply would provide energy for the entire nation for several years. A large company wants to drill for the oil and pipe it across the desert to a refinery in New Mexico. The oil deposits are under government-owned land. Ordinarily the land could be leased to the oil company and the oil could be removed without controversy. However, the land is on the verge of being declared a wilderness area. Several environmental groups have worked for years to have the area declared a National Wilderness area. The regulations governing a wilderness area will prohibit the drilling for oil. Many people see a source of cheap energy that will last for years and create many jobs, whereas many others see the area as pristine wilderness that should not be disturbed for any reason. What should the government do?

Again, divide into two groups and choose a side. One side should give reasons and arguments for continuing with the operation, and the other side should provide reasons and arguments for canceling the operation to save the wilderness area. Think through each argument and list them below.

Reason or Argument 1

Reason or Argument 2

Reason or Argument 3

Reason or Argument 4

Reason or Argument 5

EXERCISE 3 I HAD IT FIRST

A landowner has a tract of timber he wishes to cut. The trees have been growing for about 60 years and are quite valuable. The terrain is mostly rolling hills, but a few of the hills are steep. There should be no severe adverse impact on the environment if proper harvesting techniques are followed. The state government has guidelines called "Best Management Practices" that provide rules for harvesting and will protect the environment. These guidelines are suggestions and are not required by law.

The landowner and logger have no problem following the guidelines except for one rule. There is a small creek running through the timber tract. The "Best Management Practices" guidelines provide that no trees should be cut within 50 feet of any stream. Near the bank of the creek are several very large oak trees that could be sawn into valuable furniture lumber. The landowner and the logger believe that there is no environmental danger in cutting the trees because there are so many smaller trees and bushes growing on the bank. An environmental protection group believes differently. This group says that the state guidelines should be strictly followed. In fact, the environmental group is lobbying to have the "Best Management Practices" made law. The landowner believes he should be able to make judgment decisions regarding his own property. What should the government do?

Again, divide into two groups and choose a side. One side should give reasons and arguments for leaving the guidelines as suggestions, and the other side should provide reasons and arguments for passing the law to make the guidelines mandatory in order to protect riparian zones and all logging areas. Think through each argument and list them below.

Reason or Argument 1

Reason or Argument 2

Reason or Argument 3

Reason or Argument 4

Reason or Argument 5

EXERCISE 4 WANTED—JOBS

A large timber company in Oregon has just received a request from a Japanese construction company for a million board feet of lumber. This is received as good news to the community where the logs will be harvested and cut into lumber. The economic benefits will have a positive effect on all the people in the community.

It is then discovered that the Japanese company wants the lumber shipped as unprocessed logs. They want the jobs involved with processing the logs into lumber to go to Japanese people. The owners of the timber company want to sell the logs because of the profit involved. The people of the community want the company to turn down the contract and sell the timber here in the United States after the lumber has been processed. The timber company says that it needs the contract because it can make more money with this contract than by selling processed lumber here. The people of the community say they need the jobs and the money generated by the processing of the lumber. They also contend that selling unprocessed logs is selling our country's natural resources. What role does government have in this situation?

Again, divide into two groups and choose a side. One side should give reasons and arguments for accepting the contract, and the other side should provide reasons and arguments for passing a law to prevent the exporting of unprocessed logs. Think through each argument and list them below.

Reason or Argument 1

Reason or Argument 2

Reason or Argument 3

Reason or Argument 4

Reason or Argument 5

Chapter 11

WOOD CONSTRUCTION MATERIALS—*WANNA MAKE SOMETHING OF IT?*

OBJECTIVES

After completing these lab exercises, students should be able to:

- determine specific types of wood needed for different construction
- determine the strength and flexibility of different types of wood
- identify characteristics of common woods
- determine the moisture content in wood
- identify factors that affect the moisture content of wood

EXERCISE 1 WOOD PRODUCTS—WHAT'S IT TO YOU?

Background

The forest provides material to use in many different types of construction. We are surrounded with construction materials that come from trees. The type of wood material depends on the availability, price, and unique characteristics of a wood that makes it suitable for a particular job. Some woods are more attractive, some are stronger, and some are cheaper—the qualifications depend on the type of construction needed.

Materials Needed

- Access to different types of wooden objects
- References on the uses of wood

Procedures

In the following chart, examples of different objects made from wood are listed. Do research on the various uses of woods and fill in the chart with probable answers.

Object	Type of Wood	Wood Characteristics
Office desk	Oak, cherry, mahogany, walnut	Strength, stability, attractiveness
Framing for a house		
Floor of an outside deck		
Handle for a hammer		
A barrel that holds liquid		
A gun stock		
Roof decking		
Golf club head		
Baseball bat		
Living room floor		
Closet lining		
Trim around doors and windows		
Salad bowls		
Bowling alley		
Stair treads		

EXERCISE 2 WAITER, THIS IS A LITTLE DRY—DETERMINING MOISTURE CONTENT OF WOOD

Background

Lumber is one of the primary products obtained from the forest. When boards are first cut from the log, they contain a high degree of moisture that has to be removed under controlled conditions. As lumber dries, it shrinks in size and decreases in weight. If the lumber dries too rapidly, it may crack or distort by warping or cracking. Even wood that has a low moisture content is constantly swelling and shrinking as moisture is removed or absorbed from the air surrounding it. This exercise will help you understand more about the moisture content of wood.

Materials Needed

- Several blocks of wood of different types (The blocks should be the same size [about 2 inches × 1 inch × 1 inch] and of different species of wood such as cedar, pine, oak, maple, and so on. The blocks should also be of varying degrees of dryness. If possible, use some blocks from freshly sawn lumber that has not been dried or from an outdoor wood pile.)
- A pencil
- A microwave oven with a rotating tray
- Gloves
- Tongs
- A set of scales that will weigh accurately to 0.1 gram

Procedures

Using a pencil, mark a number on each block of wood and record the number in the chart provided below, along with a description of the block. For example, block #1 might be "green" wood from a freshly sawn piece of oak lumber; block #2 might be a piece of pine from the agricultural mechanics lab; and so on. Select three of the blocks, and carefully weigh them and record their weights. Then place them in the microwave oven evenly spaced along the outside edge of the carousel. Set the oven on low to medium-low power (200 to 350 watts). Heat the blocks for about 20 minutes. Using oven mitts and tongs, take the blocks out of the oven and carefully weigh the blocks and record the weights. Repeat this procedure for all of your blocks.

Once you have recorded all the data, calculate the percentage of water that had been in each sample. To do this, divide the weight of a block after it comes out of the oven by its weight before it went into the oven and multiply by 100. This will give you the moisture content of the lumber.

Hypothesis

After collecting your samples of wood, take a close look at the different samples. Determine which sample of wood you think will have the highest moisture content and write a hypothesis reflecting this choice.

____

____

____

Observations

Block Number	Description	Weight before Drying	Weight after Drying	Percent Moisture Content

Block Number	Description	Weight before Drying	Weight after Drying	Percent Moisture Content

Conclusions

1. For an added bonus, find out what the ideal moisture content is for lumber used in framing.

____

____

2. Determine the ideal moisture content for lumber used in building furniture. Why is there a difference in the desired moisture content?

____

____

3. After reviewing your observations, accept or reject your hypothesis. Explain.

____

____

____

EXERCISE 3 WOOD STRENGTH

Background

Woods vary in their flexibility and strength. A particular application might call for a wood to be flexible, whereas another might require wood that is strong and unbending. A third possibility would be an application that requires wood that is both flexible and strong. **Flexibility** refers to the ability of wood to bend without breaking, and **strength** refers to the ability of the wood to withstand bending and/or breaking.

Materials Needed

- 2 triangular blocks approximately 2 inches on a side and 3 inches long
- Ruler or yard stick
- A pencil
- Several strips of different types of wood 3/4 inches thick × 2 inches wide × 2 feet long (suggested types are pine, oak, hickory, maple, oriented strand board, particleboard, plywood, and so on)
- Weights of 1, 2, and 3 pounds

Procedures

Step 1

Lay the triangular blocks about 20 inches apart on a strong table. Put a strip of wood across the blocks, as shown in the diagram in Figure 11–1. Take your yard stick and measure the distance between the bottom of the wood and the table. Record your measurements in the chart provided on page 95.

Step 2

Place each weight on the center of the strip, beginning with the 1-pound weight, and then adding the 2-pound weight, and finally the 3-pound weight. As the weights are added, measure and record the distance between the bottom of the strip of wood and the table. Repeat this process with each strip of wood.

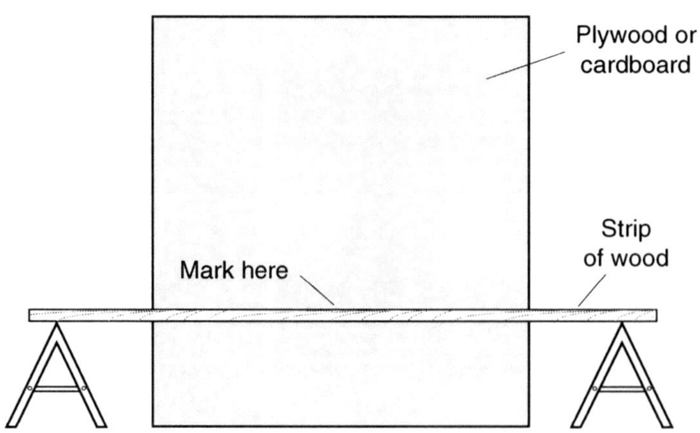

Figure 11–1

Step 3

Next, stack three pine strips on top of the blocks and start adding weights. Record the degree of bending. Then build a beam by gluing the three strips together. After the glue has set overnight, place the beam on the blocks and add weight. Record any bending.

Hypothesis

Before you begin, examine the wood and form a hypothesis predicting which board will be the strongest and which board will be the weakest.

____

____

____

Observations

Wood Type	No Weight Inches (") between wood and table	1 lb.	2 lb. added (3 lb. total)	3 lb. added (6 lb. total)
		"	"	"
		Total Change	Total Change	Total Change
		"	"	"
		Total Change	Total Change	Total Change
		"	"	"
		Total Change	Total Change	Total Change
		"	"	"
		Total Change	Total Change	Total Change
		"	"	"
		Total Change	Total Change	Total Change
		"	"	"
		Total Change	Total Change	Total Change

Conclusions

1. Which wood strips are the most flexible? Which are the strongest?

____

____

____

2. What causes the difference in the strength of the beam as opposed to that of the three pieces stacked together?

3. After reviewing your data, accept or reject your hypothesis. Explain.

Chapter 12

SPECIALTY FOREST PRODUCTS—*YOU ARE SPECIAL*

OBJECTIVES

After completing these lab exercises, students should be able to:

- discuss the steps in the paper-making process
- create paper from recycled material
- explain the various materials used for making paper

BACKGROUND

Many products other than lumber come from the forest. One primary product is paper. In the following exercises, you will make paper from a variety of materials.

EXERCISE 1 FOLLOW THE RECIPE—MAKING RECYCLED PAPER

Materials Needed

- Scrap paper (Junk mail works best. Avoid slick, glossy paper or newsprint.)
- Cornstarch
- A large coffee can
- A blender
- A small frame 1 foot × 1 foot (An inexpensive picture frame works well.)
- 2 fine screens of approximately the same size; one screen must fit inside the frame
- A flat block of wood 6 inches × 6 inches
- An old towel
- A large tub (about 2 feet × 18 inches × 1 foot deep) of clean water
- 1 cup of drier lint
- ½ cup of dried flowers
- Other types of fiber you think might make paper

Procedures

Step 1

Tear the scrap paper into strips about 1 inch wide and 2 inches long. Put approximately three cups of the paper scraps into the coffee can. Next, put about three cups of warm water into the can with the paper and let it sit for several hours or overnight.

Step 2

After the paper has soaked in the water, pour the mixture into the blender and add about a tablespoonful of cornstarch. Blend slowly at first, then work up through the speeds and liquefy for about 30 seconds.

Step 3

After blending, pour the mixture into the clean water and mix it thoroughly. Immediately place the frame with the screen bottom into the mixture at an angle while the particles are suspended. Move the screen to the bottom and slowly move it toward the top using a sifting motion. A flat layer of the mixture should appear on the bottom of the screen.

Step 4

Place the screen on sticks across the tub and allow the water to drain. Then place the frame on an old absorbent towel and place the loose screen on top of the mixture. Use the block of wood to press the excess moisture from the sheet sandwiched between the screens.

Step 5

Remove the block and the loose screen and place the screen frame with the sheet in a warm place to dry. When the sheet is dry, remove it from the screen and use it as you would use any paper.

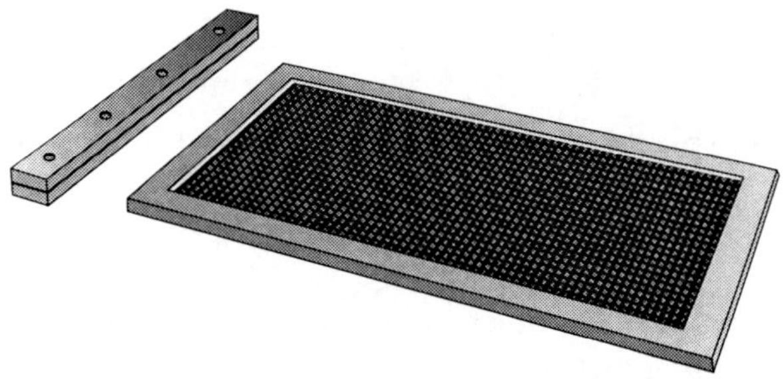

Figure 12–1

Step 6

Make another batch of pulp using other recycled materials. Tear the scrap paper into strips about 1 inch wide and 2 inches long. Put approximately two cups of the paper scraps and one cup dryer lint into the coffee can. Next, put about three cups of warm water into the can with the paper and let it sit for several hours or overnight.

Step 7

After the paper has soaked in the water, pour the mixture into the blender and add about a tablespoonful of cornstarch. Blend slowly at first, then work up through the speeds and liquefy for about 30 seconds. If the resulting paper sheet tears, try adding more cornstarch.

Step 8

Try adding some flowers, leaves, or glitter to make the paper more attractive. Cotton fibers will make a nice addition to your paper. Use your imagination and experiment with different fibers to make paper. Create a chart of the different ingredients and describe the differences in texture, strength, and other characteristics.

Observations

Materials Used/Paper Name	Appearance/Texture	Strength/Usability Good, Fair, or Poor

EXERCISE 2 MEDIUM-RARE PLEASE—MAKING PAPER FROM RAW MATERIAL

Materials Needed

- The same materials used in Exercise 1, except for the shredded paper
- 3 cups of bleach
- 3 cups of soft wood shavings
- A stainless steel pot that will hold 2 quarts

Procedures

Safety Caution

The use of the bleach and oven should be left for the instructor to do. You may watch, but be sure to stand back and wear eye protection. Perform this lab in a well-ventilated area or outdoors. Gasses may be harmful.

Step 1

Place the wood shavings and the bleach in the pot and place the mixture in an oven. Heat to about 50°C for about 24 hours. Remove the pot from the oven and check to see if the fibers have separated. If they are not separated, put the pot back in the oven and heat for about 2 more hours.

Step 2

When the fibers are loose and pliable, drain off the bleach and wash the fibers thoroughly in clean water. You will probably need to go through several rinses, using a fine mesh screen as a strainer.

Step 3

Place the fibers and warm water in the blender just as you did in Exercise 1. Then continue with the process as you did with the recycled paper. Compare the paper made from recycled materials with the paper made from raw materials.

Step 4

In small groups, use the text and the Internet to research how commercial paper producers make paper from wood. Find out what chemicals are used to whiten paper and what steps are taken to neutralize toxins left in the water. Compare and contrast what steps were taken in this lab to the procedures used in a commercial plant. Appoint a spokesperson to report your findings to the class.

Notes:

EXERCISE 3 FIRE, FIRE, FIRE—FIREWOOD HEAT POTENTIAL

Background

All trees can be burned to produce heat, but some have the ability to produce more heat than others. Much of this has to do with the density of the wood, which relates to the number of cells found per cubic inch. More cells per unit of volume create more density, which creates more BTUs (British thermal units) of heat energy. Other factors, such as the amount of turpines found in the sap, have some impact as well and increases sparks during combustion. All fuelwood needs to be dried, preferably by storing in a fairly dry environment for up to a year. This reduces the creosote that may be deposited on chimney walls as the wood burns and reduces the chance of dangerous chimney fires. Chimneys should be cleaned by a professional each year to prevent this hazardous condition.

Procedures

Data on several types of fuelwoods have been collected and are displayed in the following chart. Analyze this chart and find the information asked for below.

Using the text and the information provided in the background and by interpreting the data, complete the chart by determining the heat output potential of the fuelwoods listed.

Species	Weight (lb./cord)	Coaling Quality	Spark Output	Heat Potential
Alder	2,500	Good	Medium	
Apple	4,400	High	Low	
Beech	3,700	High	Low	
Western Red Cedar	2,100	Low	High	
Cherry	3,200	High	Low	
Cottonwood	2,100	Low	Medium	
Elm	2,200	Good	Low	
Douglas Fir	2,900	Good	Medium	
Hemlock	2,700	Low	Low	
Western Larch	3,300	Medium	High	
Black Locust	3,800	High	Low	
Sugar Maple	3,600	High	Low	
Red Maple	3,200	High	Low	
Red Oak	3,700	High	Low	
Ponderosa Pine	2,200	Medium	Medium	
White Pine	2,200	Medium	Medium	
Redwood	2,400	Low	High	
Norway Spruce	2,200	Low	High	
Willow	2,500	Low	Medium	

Conclusions

1. Explain the relationship among the weight of wood, density of wood, spark production, and heat potential.

____

____

____

____

EXERCISE 4 BRRRRRR; STOKE THE FIRE—EXPERIMENTING WITH HEAT POTENTIAL

Materials Needed

- Log samples of different species of wood from the table in Exercise 4 (enough of each kind to build a camp fire-sized fire)
- A safe area for building small fires
- Noncontact infrared laser thermometer
- Three large rocks (three inches in diameter) for each species of wood

Hypothesis

Examine the wood samples provided by your instructor. Form a hypothesis that predicts which wood will burn the hottest and which wood will put off the least amount of heat.

____

____

____

Procedures

Step 1

In a safe designated area, build small a campfire around each stone. All fires will be made with a different species of wood. Place the second rocks one foot from the fires; the third rocks, three feet away from the fires (see Figure 12–2).

Figure 12–2

Step 2

After the fires have been burning well for 10 minutes, measure the temperatures of each rock with the infrared laser. Standing at a 90-degree angle from the rocks (as seen in Figure 12–2), point the laser at each rock and take a reading. Record your data on the chart provided on page 102.

Step 3

Take two more sets of readings after 20 minutes of burn time. Record your findings on the chart and then analyze your data.

Observations

Wood Type	10-Minute Reading			20-Minute Reading		
	Base	1'	3'	Base	1'	3'

Conclusions

Using the chart in Exercise 3 and the data from your fire experiments, answer the following questions.

1. Looking at the chart in Exercise 3, which wood species would you expect to burn the hottest?

____

____

2. In the class experiments, which wood burned the hottest? Which wood burned the coolest? Looking at the chart in Exercise 3, were you surprised by your findings?

____

____

3. Write a statement accepting or rejecting your hypothesis. Explain.

____

____

____

Chapter 13

PLANTATION PRODUCTS AND PRACTICES—*WOOD YOU GET TO WORK?*

OBJECTIVES

After completing these lab exercises, students should be able to:

- discuss the proper method of transplanting seedlings
- construct a dibble to use in transplanting
- determine the best species of tree for a particular area
- determine how many seedlings are needed for a particular plot of land
- transplant seedlings properly
- write a business plan for starting a forestry business

BACKGROUND

Although some new forests are started by directly seeding the plot, almost all commercial trees planted in a plantation begin in a nursery where seedlings are produced for transplanting. Most often these seedlings are part of a selective breeding program that produces superior plants. They may be hybrids, purebred strains, or the result of genetic engineering. Beginning with superior seedlings is extremely important in establishing profitable forest plots. Just as important is the task of properly transplanting the seedlings. No matter how good the seedlings' quality is, if they are not transplanted properly, they either will not live or will not grow properly.

There are two methods of transplanting. One method is by using a mechanical planter that is usually pulled behind a tractor. The other method is planting the seedlings by hand.

EXERCISE 1 OOPS! I MADE A DIBBLE—CONSTRUCTING A TRANSPLANTING TOOL

This exercise allows you to build your own transplanting **dibble**, or planting bar, for setting out seedlings. The materials may be found at a metal scrap yard and should be inexpensive. The dibble can be easily constructed in your agricultural mechanics lab.

Materials Needed

- *Handle:* 2 to 6½ inches, threaded ¾-inch pipes
- *Tee:* ¾-inch galvanized or black iron
- *Shaft:* 2-foot, ¾-inch galvanized or black iron pipe threaded (both ends)
- *Coupling:* ¾-inch galvanized or black iron
- *Round stock:* 4 inches long by ¾ inches diameter, threaded with pipe threads on one end
- *Foot rest:* mild steel ¼ inch thick by ¾ inches wide by 8 inches long
- *Blade:* worn plow share 3 to 4 inches wide and 10 inches long, with one side sharpened

Procedures

Step 1

Using a cutting torch or hacksaw, notch out the unthreaded end of the round stock so that it fits about 4 inches deep on the blade, as shown in the diagram in Figure 13–1 on page 106.

Step 2

Weld the round stock to the blade.

Step 3

Weld the footrest to the top of the blade.

Step 4

Using pipe wrenches, assemble the parts as shown in Figure 13–1. Be sure the handle is properly in line with the blade. A spot weld on each side of the tee and on each side of the coupling should hold the shaft and handle together in the proper position.

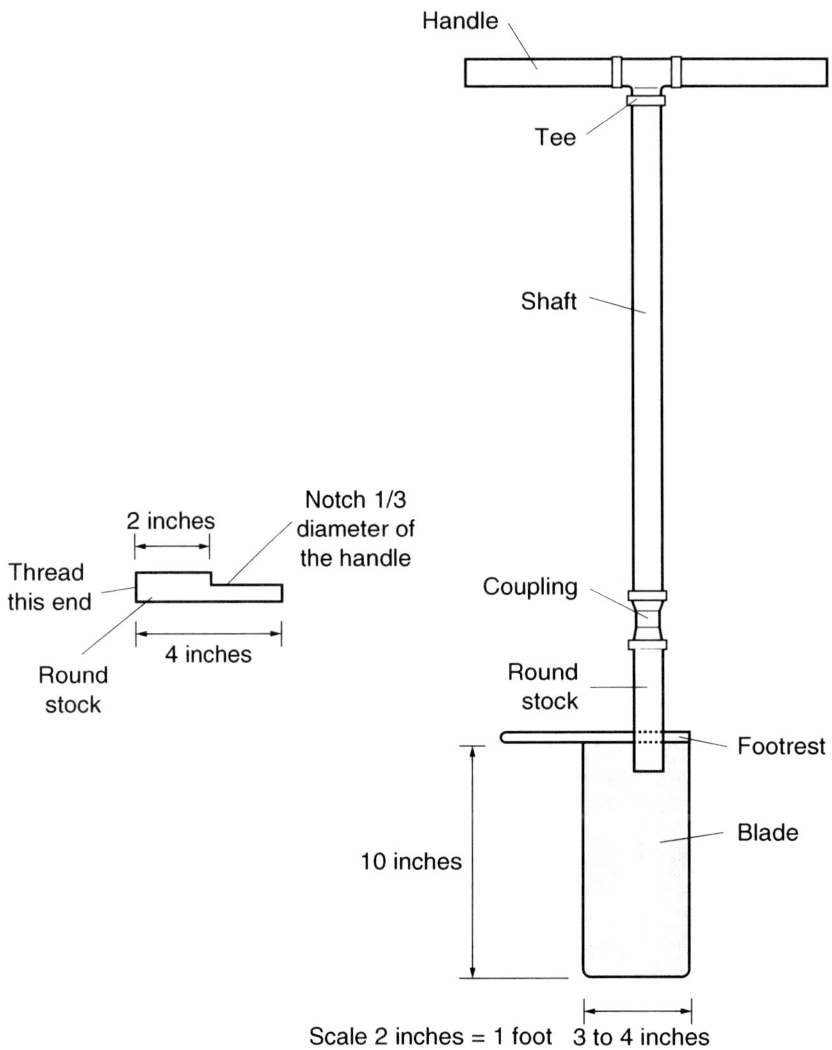

Figure 13–1

EXERCISE 2 BE-LEAF IN THE FUTURE—PLANT A TREE SEEDLING

Materials Needed

- A plot of land where you can set out a plantation of trees
- Enough seedlings to plant the area
- Wooden stakes and at least 100 feet of strong string
- A tape measure
- A planting dibble
- A bright-colored felt-tip pen

Procedures

Step 1

Locate a plot of land that should be planted in trees. Determine what the best species of trees would be to plant in the area. A local forester should be able to advise you on the types that are best suited for your area.

Step 2

Calculate the number of seedlings needed for the plot. The first step is to determine the square footage of land in the plot. Measure the sides of the land and multiply the length by the width. Check with a forester in your area and find out the recommended spacing for different species of trees. Common spacings used are 6 × 10; 6 × 12; 8 × 8; and 5 × 10. To determine the number of seedlings needed in an acre, divide 43,560 (the number of square feet in an acre) by the spacing. For example, if you use a 6 × 12 spacing, multiply 6 times 12. This gives 72. Divide 43,560 by 72, and you get 605 seedlings needed per acre. If your plot is a half acre, then divide 605 by 2.

Step 3

Once you have decided on the species and the number needed, contact your state Forestry Commission for a source of the trees. The commission may be able to arrange to have enough seedlings delivered to your school.

Step 4

When the seedlings arrive, they should be stored in a cool place and the roots kept moist (but not wet). Set them out as soon as you can because the plants will be healthier and grow better if they are not in storage very long.

Lay out the string on the ground and measure the spacing you will be using. For example, if you are using a 6 × 12 grid, mark the string every 6 feet with the felt-tip pen. Stretch the string between two wooden stakes driven in the ground and plant a seedling at every mark. When the first row is done, move the stakes at both ends exactly 10 feet from the previous row and again use the marks as spacers. By using this method, you can arrive at perfect 6 × 10 spacing.

Step 5

The seedlings should be planted using the following technique:

1. Insert the dibble straight down into the soil to the full depth of the blade and pull back on the handle to open the planting hole (see Figure 13–2). *Do not* rock the dibble back and forth, as this causes soil in the planting hole to be compacted. Compacting the soil inhibits root growth.
2. Remove the dibble and push the seedling roots deep into the planting hole (see Figure 13–3). Pull the seedling back up to the correct planting depth (the root collar should be 1–3 inches below the soil surface). Gently shake the seedling to allow the roots to straighten out. *Do not* twist or spin the seedling or leave the roots J-rooted. **J-rooted** means that the ends of the roots are pointed toward the top of the hole.

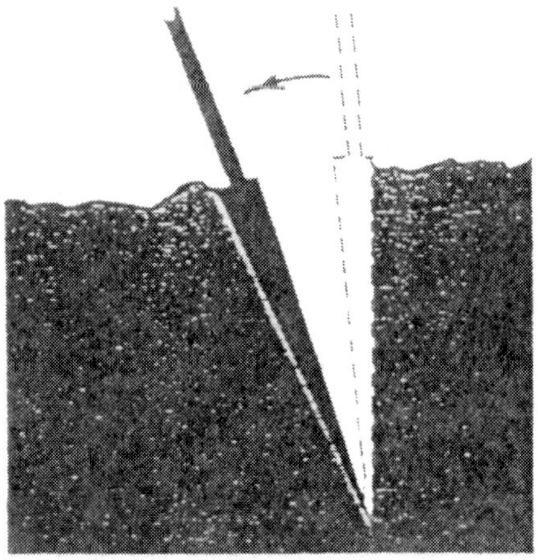

Figure 13–2 *(Courtesy Georgia Department of Education)*

Figure 13–3

3. Insert the dibble several inches in front of the seedling and push the blade halfway into the soil (see Figure 13–4). Twist and push the handle forward to close the top of the slit to hold the seedling in place.
4. Push the dibble down to the full depth of the blade (see Figure 13–5).

Figure 13–4 *(Courtesy Georgia Department of Education)*

Figure 13–5

5. Pull back on the handle to close the bottom of the planting hole. Then push forward to close the top, eliminating air pockets around the root (see Figure 13–6).
6. Remove the dibble and close and firm up the opening with your heel (see Figure 13–7). *Be careful* to avoid damaging the seedling.

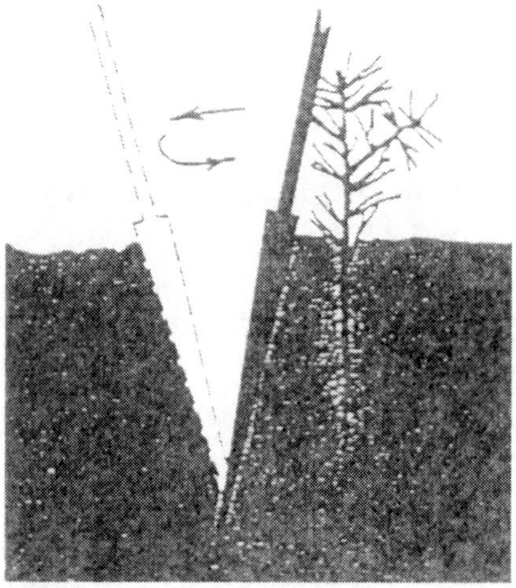

Figure 13–6 *(Courtesy Georgia Department of Education)*

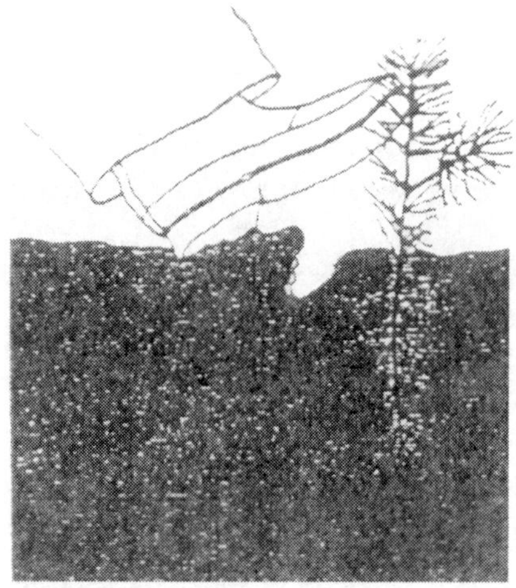

Figure 13–7

Carefully monitor the seedlings, count the number that die, and calculate the survival percentage. You should get better than an 80 percent survival rate.

EXERCISE 3 IT TAKES MONEY TO MAKE MONEY—STARTING A CHRISTMAS TREE LOT*

Background

Most of the fresh Christmas trees sold each year are raised in Christmas tree plantations. Plantation trees are generally higher in quality than wild trees and are harvested easier and with smaller costs involved. In recent years, more and more entrepreneurs have tried to earn a few extra bucks selling Christmas trees. In this lab, you will write a simple business plan for the creation and operation of a Christmas tree sales lot. If you are comfortable with your plan, don a Santa hat and sell some trees. This would make a very good SAE project or FFA chapter fund-raiser.

Procedures

Step 1

Name your business. The name should be relevant to your type of your business. It should be easy to spell and remember.

____

____

*Adapted from *8 Steps to Creating a Simple Business Plan for 2010* by Herman Drost.

Step 2

Write a mission statement. A mission statement defines your business. It will define what your business does and will describe what sets it apart from other similar businesses.

____
____
____
____

Step 3

What is your vision for your business? What do you expect it to be like in five years? Will you have more than one lot? How many employees will you have?

____
____
____
____
____

Step 4

Make some business goals and objectives for your tree lot. State clearly what you want to accomplish with your business. These goals and objectives need to be measureable and have an appropriate time line.

____
____
____
____
____
____
____
____

Step 5

Before you start a business, it is important to be able to evaluate your proposed business from a nonbiased perspective. In the chart provided, constructively answer these questions:

	Possible Issues	Resolutions to Issues
Strengths		
Weaknesses		
Possible threats		
Opportunities		

Step 6

A good action plan is vital to getting your business up and running. In this plan, you will create a marketing plan. Your potential customers need to know what your business offers, where you are located, and why they should use your business over other similar options. This is how you will sell your business.

____

____

____

____

____

____

Step 7

It's time to talk money. Every business proposal needs to include a financial plan. This is basically the projected business budget. You need to estimate your expenses and potential income. You will need to consider rent, utilities, credit card fees, equipment, payroll, taxes, and cost of your product, etc. Obviously, you want to make more than you spend. Consider all factors and determine if this is a business that will make money or lose money.

Item	Cost	Expense
Total $		

Step 8

Now that you have a plan, take a good look at it and decide if you are ready to jump in with both feet or if you need to do some more planning. If you are ready to take the leap, be sure to come back regularly and check your plan to see if you are on the right track. The business will only succeed if you keep a constant eye on the big picture. Good luck and happy money-making.

Chapter 14

URBAN FORESTRY—*TREES IN TOWN*

OBJECTIVES

After completing these lab exercises, students should be able to:

- identify sites for an urban forest
- analyze components of an urban forest development plan
- construct a plan for developing an urban forest
- present the plan to local government officials

EXERCISE 1 FAIL TO PLAN, PLAN TO FAIL—DEVELOPING A PLAN FOR AN URBAN FOREST

Background

Most communities have trees that are a valuable part of the community environment. Most often these trees have been planted for a particular purpose, such as for shade, noise control, wind control, or beauty. Whether or not the urban forest accomplishes the purpose for which it was planted depends to a large degree on the planning that goes into the development and management of the plot. The most effective and beautiful urban forests are those that have been thoroughly planned *before* the planting began. Quality planning takes a lot of time and effort, but it provides the basis for a project that can be enjoyed for generations to come.

Procedures

Obtain a map of your city, town, or community. Draw in the places where there are trees or urban forests. Select a site that will be suitable for the establishment of a new urban forest. The site does not have to be large, but there should be some reason to establish trees in that area. Using the following list of items for consideration, construct a plan for developing an urban forest. Be thorough in your analysis and planning, and be sure to answer the questions completely, so that anyone who reads the plan will comprehend what you are proposing and why.

Step 1

Location: Write a description of where the plot is to be located. Draw the area in on the map, and draw a sketch of the area that includes a description of the terrain, buildings, roads, sidewalks, streams, and any other physical attributes of the area.

Step 2

Purpose: What is the primary reason for planting the trees and developing the area? Explain reasons such as beautification, wind protection, noise reduction, shade, and so on. Be thorough in your explanation.

Step 3

Proposed Species: What types of trees do you propose? Why did you pick these particular species? Are they evergreen or deciduous, blooming or nonblooming? Are they suitable for your hardiness zone? Are these trees available in your area? Where can you purchase them?

Step 4

Added Benefits: Are there added benefits to having the trees in that location? For example, will wildlife be brought in as a result? Oaks and hickories might be a habitat for squirrels and other small animals. Mulberry trees might draw different species of birds.

Step 5

Maintenance Schedule: What will have to be done to keep the trees healthy, growing, and serving their purpose? Will they need watering, fertilizing, or pruning? Think of all the maintenance activities that need to go into managing the trees. Do not forget to include *who* will be responsible for maintenance. Consider involving the local parks department to help with the maintenance.

Step 6

Costs Involved: Make an itemized list of all the costs involved with establishing and maintaining the trees. Items such as fertilizer and water should be included, along with the cost of the trees.

Step 7

Funding Sources: Discuss how you think the project could be funded. The help of the city or county commission, civic groups, and other organizations should be explored. Also think about the possibility of your FFA chapter (of the National FFA Organization) funding the project. Consider asking for donations from local nurseries. Be sure to thank sponsors properly. A permanent sign with their name on it would be appropriate.

EXERCISE 2 GET 'ER DONE!—IMPLEMENTING THE PLAN

Background

Many forestry careers will eventually require that you present information about the industry to other people. For example, you may find yourself simply defending your profession to an acquaintance, or you may be asked to speak in front of state legislators about logging regulations or restrictions. It is important to become comfortable and confident speaking in public. The best way to do this is with the three P's—planning, practicing, and presenting. This is exactly what you will do in the lab, so take a deep breath. Close your eyes. Hum a bit if it helps you relax, and get ready to conquer the most common phobia in the world: public speaking.

Procedures

Step 1 *Planning*

After the proposal is complete, discuss what presentation format will work the best. As a class, decide if you will use a computer program like PowerPoint. Will you use an overhead projector or make a large flip chart?

Step 2

Decide who will present the information. Will all class members have an opportunity to present, or will a few of you do most of the talking?

Step 3

Break the class into groups of eight. Each group will write a 2–3 minute presentation about one of the following topics from Exercise 1: location, purpose, proposed species, added benefits, maintenance schedule, costs involved, and funding sources.

Step 4

Using the presentations from step 3, create a PowerPoint slide, poster, or overhead for your group's main point. Keep in mind that a visual should touch on the main points and not have a great amount of wording. If the class is using PowerPoint, the slides should have a common design with very few fancy swooshes and flying words. Professional presentations focus on the information, not on how well the presenter can work the computer program.

Step 5

Next, all presenters should decide in what order they will present their information and give that information to the nonpresenters. The nonpresenters (let's call you the "support team") will then organize the visual part of the presentation. Get charts in the proper order or put all of the PowerPoint slides into one presentation in the correct order.

Step 6 *Practice*

While the support team is hard at work, find an area in the classroom where you will feel comfortable talking to yourself. For the next 5–10 minutes, practice your part. When the support team is ready, practice your presentation a few times.

Step 7 *Present*

Your instructor or an appointed committee will invite local government officials to come to the class for the presentation. If it is not possible for them to come to you, arrange to speak at a town meeting. Present the plan as you have developed it and be ready to answer any questions. Good luck, and remember to smile.

Step 8

As soon as the plan has been approved, organize your class or FFA chapter into committees and implement the project. Be sure to include the project as part of your National Chapter Award application under the Community Development component.

EXERCISE 3 TAKING AN INVENTORY OF THE URBAN FOREST

Background

Urban forestry is becoming an increasingly important field. Many colleges and universities are offering degrees in this field, and urban foresters are in increasingly high demand. Every community has trees along its streets, in its parks, and planted in landscapes around its public buildings. The health and management of these trees must be considered by each municipality so that growth is controlled and sick or weak trees do not create public risks. The first step in managing the urban forest is to know what you have to work with. An inventory must be taken of all the trees within the jurisdiction of the community. The inventory should list not only the types of trees, but also their general size and condition.

Materials Needed

- tree identification guide
- *Introduction to Forestry Science* textbook Chapter 19

Procedures

Step 1

Using the chart provided on page 116, create an inventory of all the trees on your school grounds. Use the identification guide and text to properly identify each species of trees.

Step 2

Then expand your range and inventory the trees in local parks and in the landscapes surrounding your public buildings. If you get even more ambitious, you can inventory the trees along the public streets and rights-of-way. In larger areas, it would be more time-efficient to break into small groups and break the area into smaller areas. This is a big job, and you may be surprised by the number of trees growing in your community. When you see how many trees are living in our communities, you will understand why urban forestry is such an important field.

Observations

Common Name	Scientific Name	Estimated Height	Condition	Health	Recommendations
Example—Red Maple	*Acer rubrum*	40 feet	Overgrown	Good	Needs pruning

Step 3

As a class, combine your data. Keep the school information separate from any community inventory you may have done.

Step 4

Decide as a class what to do with the information you collected. Was it just fun to get outside and look at trees and that is as far as it goes? Do you want to pass the information on to a parks department or school staff? After you have decided what to do with your inventory, follow through with that decision.

Volunteering to help with some maintenance could be a great service project or possible fund-raiser. You could even involve the community by asking for volunteers. Organizing and supervising this large scale project would make a great SAE or Eagle Scout project.

Chapter 15

SPACE-AGE FOREST TECHNOLOGIES—*BEAM ME UP, SCOTTY!*

OBJECTIVES

After completing these lab exercises, students should be able to:

- define terms used with computers and computer technology
- demonstrate an understanding of what is involved in managing a forest

EXERCISE 1 IT'S VIRTUALLY A FOREST—COMPUTER SIMULATION OF THE FOREST INDUSTRY

Background

The best way to learn something is by doing it. We cannot spend the next year in a forest, but we can spend a day or two online simulating a year in the forest. FORESTIA is a free educational game created by Creo Game for Science. In this virtual world, you create and are responsible for managing a forest in a way that satisfies all users' needs fairly. You will be given advice from guide characters throughout the game. They will help you take inventory, teach you how to harvest trees and to fight fires, and help you choose the best area for a protected area.

Procedures

Step 1

Log onto http://www.gameforscience.ca/games.php and set up an account. Use good judgment when giving any personal information over the Internet. Do not use your name, address, birthday, etc.

Step 2

Take advantage of the tutorials until you feel comfortable going solo. Have fun as you get an almost first-hand virtual education.

Step 3

Keep notes on major successes and setbacks.

Conclusions

Now that you have experienced managing a forest, summarize your experience in a short 1–2 page report. Use separate sheets for this report. Be sure to include your successes and your challenges. Briefly discuses daily activities where you needed to use a computer device. Comment on the importance of computer-based technology in today's modern forests.

EXERCISE 2 A-TO-Z TO DOWNLOAD WHAT?—COMPUTER TERMINOLOGY

Background

Computers have completely changed the way we find, sort, store, and use information. The technology is so rapidly changing that it is almost impossible to stay updated with equipment, technology, and information. The Internet continues to grow at a phenomenal rate, and the amount of information we have access to is almost beyond comprehension. Many technologies based on computers have emerged and are being used in the forest industry. The following exercise uses a fun way to learn some of the vocabulary associated with computers and computer technology.

Procedures

Use your textbook to solve the following crossword puzzle. Keep trying until you are able to fill in all of the blanks.

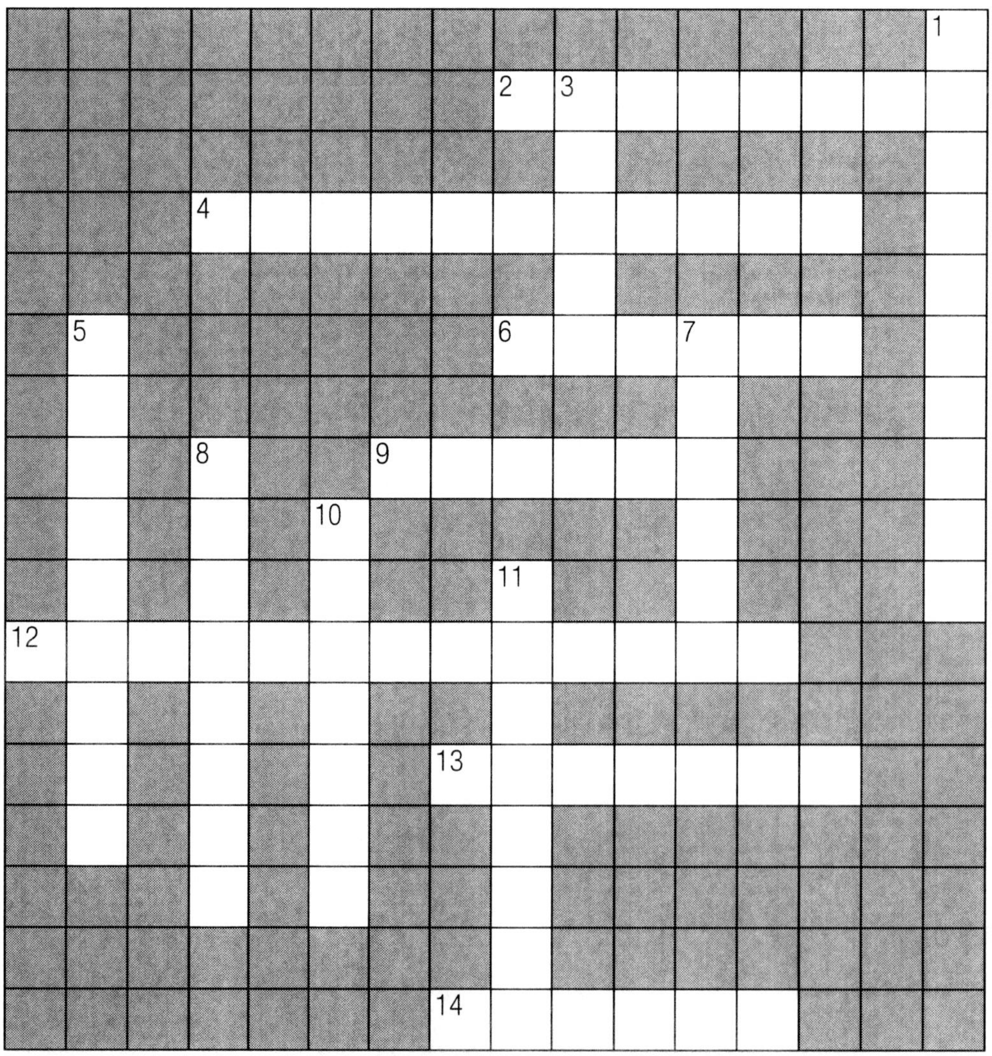

ACROSS

2 Earth ________ satellites transmit different types of data to earth
4 A map that shows elevation
6 ________ positioning systems allow you to find your location
9 Forest ________ models estimate forest yields
12 Instruments used to analyze satellite photos
13 A delivery system linking two or more computers
14 Photography taken from a plane or satellite

DOWN

1 ________ information system combines data with mapping to evaluate forest needs
3 Method of sending electronic messages
5 Forest ________ is an estimation of volume of timber at a given time
7 A computer program that deals with simulated forest fires
8 Photography sensitive to different wavelengths of light
10 Organized commands used by computers to perform calculations
11 Venue of information a person or organization has available on the Web

Chapter 16

PHYSIOLOGY OF TREES—*WHAT MAKES THIS THING TICK?*

OBJECTIVES

After completing these lab exercises, students should be able to:

- use an increment borer
- determine the age of a growing tree
- determine the growth rate of a live tree
- draw conclusions based on increment bores

EXERCISE 1 YOU'RE GETTING SO TALL!—TREE GROWTH

Background

Trees generally add a growth ring each year. This means that you can estimate the age of a harvested tree by counting the annual growth rings on the stump. Foresters often need to know the age of a growing tree in order to make management decisions. A tree with a small diameter may in fact be older than a tree with a large diameter. This may be the result of better growing conditions, better genetics of the tree, or other factors. Knowing the true age of a group of trees helps foresters to make decisions regarding thinning, clear-cutting, or other management practices.

An instrument called an **increment borer** can be used to take a core sample of a tree (see Figure 16–1). The increment borer consists of a hollow tube attached to a T-shaped handle. The tube is designed to be screwed into the trunk of a tree and to cut a thin sample of its cross section. A thin metal device is inserted into the tube and the core is extracted.

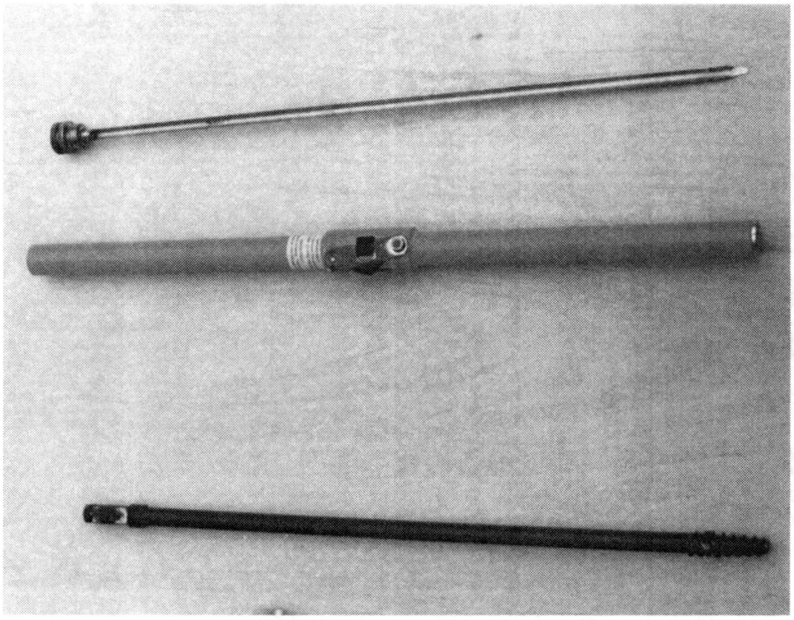

Figure 16–1

Materials Needed

- Pliable measuring tape
- Increment borer

Hypothesis

Write a statement predicting whether or not proximity to other trees affects the width of their growth rings.

____

____

____

Procedures

Step 1

To use the increment borer, remove the center "saber" and press the sharp end on the trunk of the tree about chest high (see Figure 16–2). Angle the instrument slightly upward and turn the handle clockwise. The cutting edge should enter the wood and continue going deeper until the cutting end reaches the center of the tree. Turn the handle about one-fourth of a turn counterclockwise (see Figure 16–3). When the borer has reached the center of the tree, carefully insert the thin saber and remove the core. Be careful not to break the core (see Figure 16–4 and Figure 16–5).

Figure 16–2

Figure 16–3

Figure 16–4

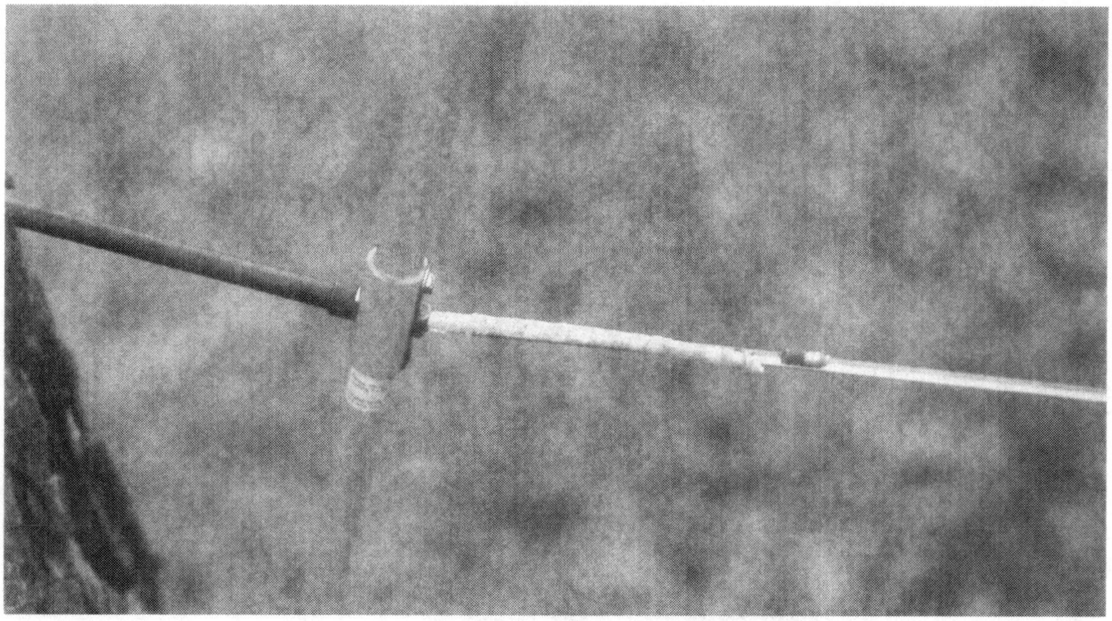

Figure 16–5

Step 2

Examine the core to determine the age of the tree. Count the annual rings to the center of the core. The center appears as an oval in the middle of the core (see Figure 16–6). What can you tell from the core? What does it mean when some rings are very close together and some are wide apart? Record your data.

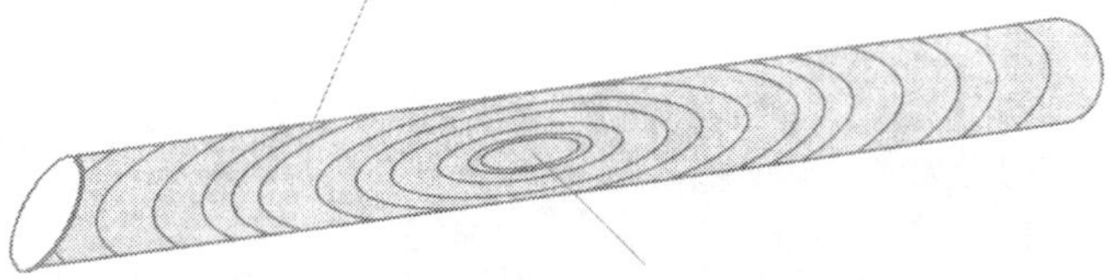

Figure 16–6

After making a careful examination of the core, carefully place it back into the tree from which it came (see Figure 16–7). Take a breast height diameter (dbh) measurement with the measuring tape.

Figure 16–7

Step 3

Select 10 trees of differing diameters and take a core from each. Record the data collected in the following chart.

Observations

Tree Number	Diameter[a]	Age[b]	Nearest Tree[c]

[a]In inches [b]Number of annual rings [c]In feet

Conclusions

Write a brief statement accepting or rejecting your hypothesis. Explain.

____

____

____

Write a short paper describing this experiment. You will be expected to use proper scientific format. An explanation of this format can be found in Exercise 3. You will need to include a reference from a published scientific article, textbook, or reliable Internet source that had similar results. People who write scientific papers use quotes from other people's research to back up their own work. You will need to cite all of your references in APA style (used in all scientific research), both here and in the text of your paper.

EXERCISE 2 TREE BLING—UNDERSTANDING GROWTH RINGS

Background

Trees actually produce what appear to be two visible growth rings per year. In actuality, the tree trunk adds new cells to its outer layer under the bark as the cambium cells divide. These new cells become wood. In the spring, the cells produced are usually much larger that those produced at other times of the year, and they appear lighter in color because of the large spaces between their walls. As the tree grows through the heat of the summer, the cells become smaller and the higher concentration of cells appears darker. As fall approaches, temperatures drop and growing conditions become less desirable; the cells produced then are smaller and appear darker still, so that the appearance of a dark ring can be noted.

Materials Needed

- Cross-sectioned tree rounds about 1 inch thick from a variety of trees
- Hand lenses, magnifying glasses, or dissection scopes

Hypothesis

Think about what factors may affect the growth rings of trees. Write a statement predicting what you may learn about a tree from its rings.

____

____

____

Procedures

Step 1

Select five tree rounds. Count the number of light rings on each round, then count the number of dark rings. How do the numbers compare?

____

____

____

Step 2

After determining the age of each round, look for differences in growth that would determine good growing years and poor growing years. Research the weather history for those years in your area and see if a correlation exists.

Step 3

Look for defects in the wood or growth patterns that would indicate fire damage, insect damage, or mechanical damage to the tree as it grew. Research the history of the area to see if fire, insect infestations, or logging activity may have caused these defects.

Step 4

If all of the tree samples are from the same area, look for patterns and possible historical explanations. For example, when the tree rings from most of the samples have a narrow spacing in year 7, look for an overall trend in precipitation for that year. Enter your findings here.

Step 5

Research human history over the life of the tree rounds and list some of the important events that occurred during the trees' life span.

Observations

In the space provided, draw each tree round. If the round has more than 10 rings, there is no need to draw each ring. Instead, draw those rings that appear out of the ordinary. Be sure to label these significant lines with ring number.

Ring A
Age ____

Ring B
Age ___

Ring C
Age ____

Ring D
Age ___

Ring E
Age ____

Conclusions

1. What kinds of events affected the tree rings on your samples?

____

____

____

2. Based on the research you did, are you able to accept your hypothesis? Explain.

____

____

____

EXERCISE 3 I'M STOKED TO WRITE A SCIENTIFIC PAPER

Background

A true scientific paper is formatted in a way that would make your English teacher cringe. It will have blunt breaks, few transition sentences, no flowery fluff, and bold labels throughout. This type of paper is not meant to persuade, demonstrate, or describe. It is a simple way to transfer information in a universally understood manner. Many forestry students will go on to study the sciences in college. Having the experience writing this kind of paper will give you the upper hand.

Procedures

Step 1

Run your experiment and keep good records. Analyze your data and begin to form conclusions in your mind. Review similar work from others in the scientific community and warm up your typing fingers, because it is time to start writing.

Step 2

Your paper will consist of five sections: abstract, introduction, methods and materials, results, tables and figures, discussion, and works cited. This is the order they will appear in your paper, but is not always the order in which you write the sections. For example, many times you will create your tables and figures before you write your results. It is nice to have your data organized, so you can explain it easily.

When you write your paper, each section will be typed in bold letters as a heading. Be sure to only include information pertinent to each section under its given heading. For example, in the results section, talk only about your findings. Do not give a reason for your observations in the results section; this information will have a place later in the paper.

Step 3

Abstract. An **abstract** is a short summary of the entire paper. It should introduce your original question. Quickly mention your experiment, the results you found, and a few sentences involving your conclusions. The abstract is found at the beginning of your paper, but it should be the last part you write.

Introduction. The **introduction** is the section of your paper where you describe the question tested by the experiment. Explain, in detail, why finding an answer to your question is interesting or important.

Materials and Methods. The **materials and methods** section should explain in detail what was actually done. Describe how you set up your experiment, and list all of the items you used. The details are important, because others may want to replicate your work. (This is done regularly in the science field.) If one small detail is left out, it may not be considered correct. ***Note:*** Do not include any results or discussion here.

Results. In the **results** section, you will provide the information learned in this experiment. Explain what you observed. Report any information that you discovered as a result of your experiment. Based on your findings, you will include a statement accepting or rejecting your hypothesis. You need to explain all of your graphs, charts, and pictures and discuss each one. You will not show your data in this section; they will be in the tables and figures section. In order for the reader to know what piece of information you are talking about, you will need to label all of your tables and figures. In the body of your paper, you will need to refer to each specific table or figure as you write about the results it represents. Here is an example. "As you can see in Figure 1, last Saturday's temperature was much cooler than Friday's."

Tables and Figures. The **tables and figures** section will contain only your pictures (called "figures"), charts, and graphs (called "tables"). Each one should have a short description and a label. For example, a copy of the first spreadsheet you discuss may be labeled "Table 1" and have a description of "Tree Data Collected."

Discussion. In the **discussion** section, you will explain your results. This is the "Why" part of your paper.

Do not describe the results again; instead, explain why you think you got the results you did. This is the section of your paper where you will include the work of others who had similar results to yours. You may take direct quotes from their writing, but you must give proper credit. Lastly, at the end of your paper, answer the question: Was your hypothesis correct? Can you accept it or have you rejected it? Explain why.

Note: If your hypothesis was not supported by your experiment, do not think of it as a failure. Many amazing discoveries were found as a nonsupported hypothesis (think of penicillin).

Works Cited. The **works cited** section of your paper is where you give credit where credit is due. You are already familiar with a bibliography from English class, and the works cited section is similar to that. Most scientific papers expect references to be cited in APA format. Instructions for citing in this style can be found at: http://www.apastyle.org/learn/faqs/cite-newspaper.aspx
For example, if I had used the text from this class as a reference, this is how I would cite it:

Burton, L. D. (2012). *Introduction to forestry science* (3rd ed.). Clifton Park, NY: Thomson Delmar Learning.

Chapter 17

FOREST ECOLOGY—*YOU'VE GOT A LOT GOING ON*

OBJECTIVES

After completing these lab exercises, students should be able to:

- find organisms in the forest
- analyze the place of specific organisms in the forest ecology
- describe the characteristics of a forest soil
- inventory soil organisms

BACKGROUND

Many types of organisms live in the forest. Some are helpful, some are harmful, and many exist without helping or harming the ecology of the forest. In many ways, these organisms depend on each other for existence. One factor they all have in common is that they depend, either directly or indirectly, on the soil. Even the higher-order animals such as deer and foxes depend on the soil to produce plants. Whatever food chain and however many organisms are on the chain, the beginning is always the soil. The only exception is aquatic animals that live all their lives in the water.

When you take a casual walk through a wooded area, you may not be aware of the many organisms that live in the area. A closer examination will reveal the different types of organisms that help make up the ecology of the forest. The following exercises are designed to help you discover the organisms in the forest and their dependency on the soil.

EXERCISE 1 HAVE WE MET?—IDENTIFICATION OF WILDLIFE AND PLANTS

Materials Needed

- Large roll of string to mark off area
- Pen or pencil
- Paper to record data

Procedures

Step 1

Locate a wooded area near your school. Pace off an area 100 by 100 steps. Locate any sign of animal life in the area. Signs might include the following:

- Leaves or other plant parts that are partially eaten
- Tracks
- Tree bark that has been disturbed
- Holes dug in the ground
- Nests
- Droppings
- Disturbed plants

Proceed slowly and cover all parts of the ground. Be sure to look up, down, and at eye level. There may be signs that are not listed above, so be alert and use your observation and analyzing skills. Remember that insects are living organisms too! Record your findings on the chart provided below.

Hypothesis

What kinds of animals do you expect to find evidence of? Write your answer in the form of a hypothesis.

____

____

Observations

Type of Sign	Animal Identification	Ecological Effect[a]

[a]Here, enter "beneficial," "harmful," or "neutral."

Step 2

Draw a map of the area you have examined, and indicate where you found signs of animal life and the type of animal you think made each sign. The sketch in Figure 17–1 is an example.

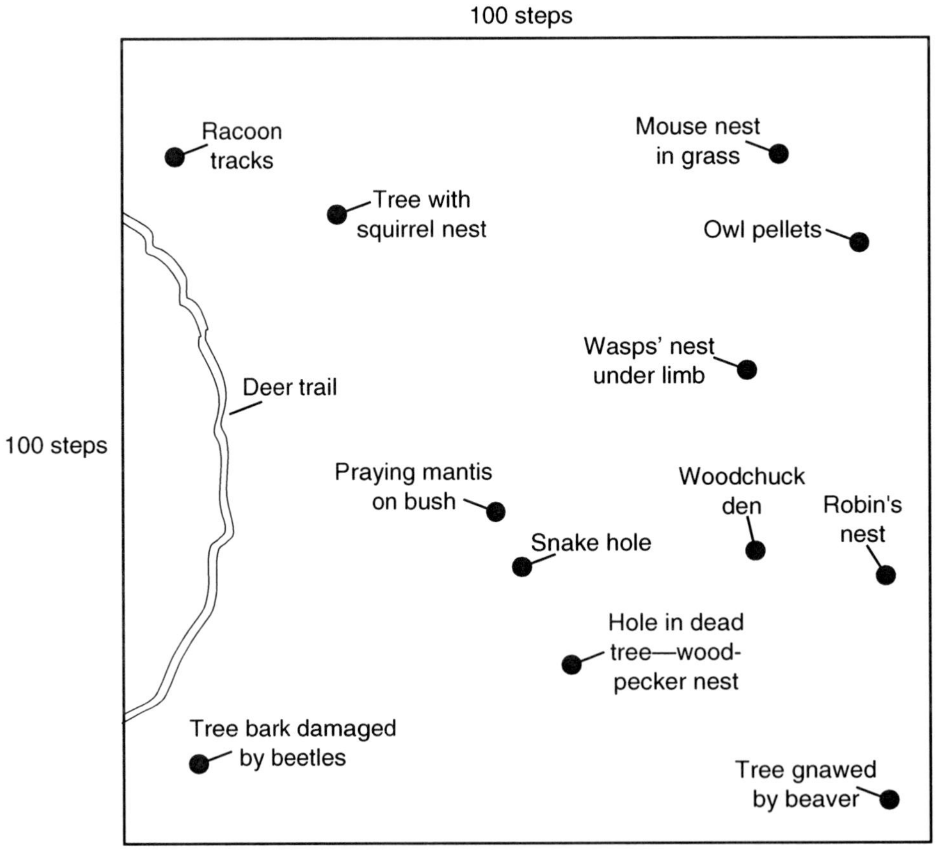

Figure 17–1 Map of Where Animal Signs Were Found

Step 3

Using the same plot of land, examine the area again and identify all the plants. Using the following chart, record the number of occurrences of each plant and the benefit that each plant might provide for the ecology.

Observations

Plant	Number of Occurrences	Ecological Benefit[a]

[a]Here, enter "beneficial," "harmful," or "neutral."

Conclusions

1. Once you have inventoried and mapped all the plants and animals observed in the plot, in the space below, write a paragraph about the interrelationship or dependency of the plants and animals in the area. Why are those particular plants and animals in that particular area?

   ____

   ____

2. Refer back to your hypothesis. Write a brief statement accepting or rejecting your hypothesis. Explain.

   ____

   ____

   ____

EXERCISE 2 LET'S GET THIS PARTY STARTED!—LIFE IN THE SOIL

Background

Soil is one of the world's most important natural resources. Nutrients for all living things are derived from the soil. It is also the home of hundreds of thousands of unique organisms. You have probably not put much thought into the organisms that live in the soil, yet these anonymous little creatures perform one of the most important functions on the planet: decomposition. Most of these small organisms remain unseen to the naked eye.

Bacteria, fungi, and small insects are the unsung heroes of the whole "Go Green" trend. They recycle more dead organic matter than you or I could ever toss into the recycling bin. In this lab, you will be able to uncover and observe some of the larger decomposers.

Materials Needed

- Shovel
- Plastic bag

- Heat lamp
- Large funnel
- Plastic bucket

Hypothesis

Predict what kinds of organisms you will find in the forest soil; form your prediction into a hypothesis.

Procedures

Step 1

When the weather is warm, go into a wooded area and locate an area covered with duff. **Duff** is the dead, nondecayed material on the forest floor. Without disturbing the duff, carefully insert the shovel into the ground and dig down about 6 inches. Quickly, without losing any material, place the duff and soil into the plastic bag. Dig out an area about 1 foot square and 6 inches deep.

Step 2

In the classroom, place the funnel across the top of the bucket with the spout down. Carefully place the material from the bag into the open end of the funnel. Make sure all of it comes out of the bag.

Step 3

Then place the heat lamp about 8 inches from the top of the funnel (see Figure 17–2). Leave the light on for several hours, and then check the bucket.

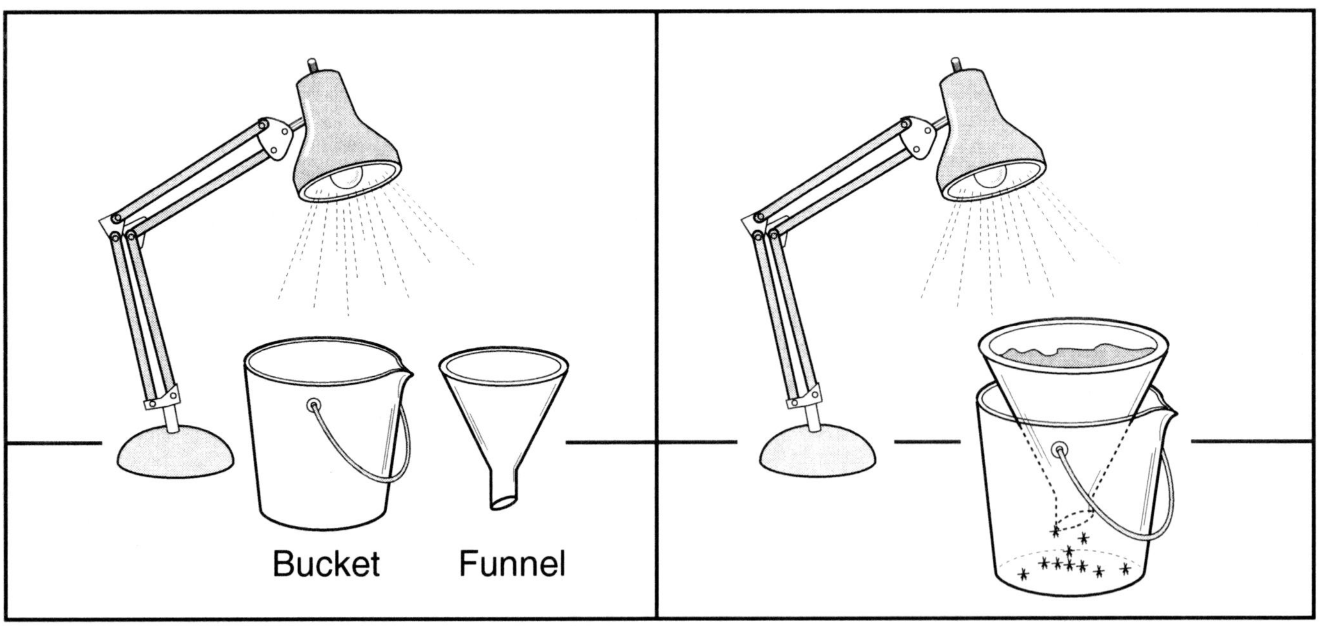

Figure 17–2

Step 4

Carefully inventory all the organisms in the bucket. Using the chart below, list the different types of organisms and the number of each type found. In the space provided, draw pictures of each organism you can see. Without referring to the identification guide, label each organism with a common name.

Type of Organism	Number of This Type of Organism

Drawings of Types of Organisms Found

Conclusions

1. Assuming that the square foot is typical of the area, calculate the number of each type of organism per acre.

2. Try to analyze how these organisms affect the ecology of the forest. What role does each organism play?

EXERCISE 3 BREAK IT ON DOWN—AEROBIC DECOMPOSITION

Background

There are many organisms in the duff on the surface of the soil. This duff is made up of leaves, needles, branches, cones, etc. These are being broken down by the soil organisms, which need food, water, and adequate temperatures to do their work.

Materials Needed

- Two plastic bottles (1 liter each) or other plastic containers such as milk cartons
- Candle
- Pliers
- Nails
- Box cutter or shears

Hypothesis

After reading the procedures below, estimate how long it will take for soil decomposers to break down newspaper strips. Form your estimate into a hypothesis.

Procedures

Step 1

Clean out a plastic bottle and cut off the top. Heat up a nail by holding it with the pliers in a candle flame. Melt several small holes in the plastic bottle to provide airflow.

Step 2

Add organic material collected from the forest floor to this bottle, as well as to a bottle with no air holes. If refrigeration is available, you may prepare another bottle as in step 1 and place it in the refrigerator.

Step 3

Make note of the condition of the duff as you add it to the bottle. How intact are the leaves, branches, needles, etc.?

Describe the duff's condition: ______________________________

____

Step 4

Let the organic materials decay for 1 week and then poor them out onto a newspaper and examine them again. How much change has occurred? Describe their condition.

____

____

____

Observations

1. If you have any bottles in the refrigerator, follow the same procedure and compare the results you observe. Describe their condition.

   ____

   ____

   ____

2. Compare the condition of the bottles stored in the refrigerator to those stored at room temperature.

   ____

   ____

   ____

3. Refer back to your hypothesis. Can you accept or reject your hypothesis? Make a comment on how temperature affects decomposition time.

   ____

   ____

   ____

Chapter 18

DISEASES AND PESTS OF TREES—*YOU BUG ME*

OBJECTIVES

After completing this lab exercise, students should be able to:

- identify insect pests that attack trees
- identify diseases that attack trees
- describe control measures used to fight diseases and pests

EXERCISE 1 WHAT'S WRONG WITH YOU?—IDENTIFICATION OF DISEASES AND PESTS

Background

Insects come in a wide variety. Some are tiny, whereas some are relatively large. Some have a hard shell, and others are very soft. It is estimated that there are as many as 30 million different species of living insects in the world. Scientists predict there are 10 quintillion (10,000,000,000,000,000,000) individual insects alive on the earth at any given time. That's a lot of "bugs." In an attempt to organize and understand insects, an international classification system organizes all insects into groups according to different criteria. There are about 30 orders of insects, but only about 6 of these damage live plants. A seventh order to which termites belong is included in our list, not because termites damage living plants per se, but because of the damage they can do to dry wood and forestry products (see Figure 18–1).

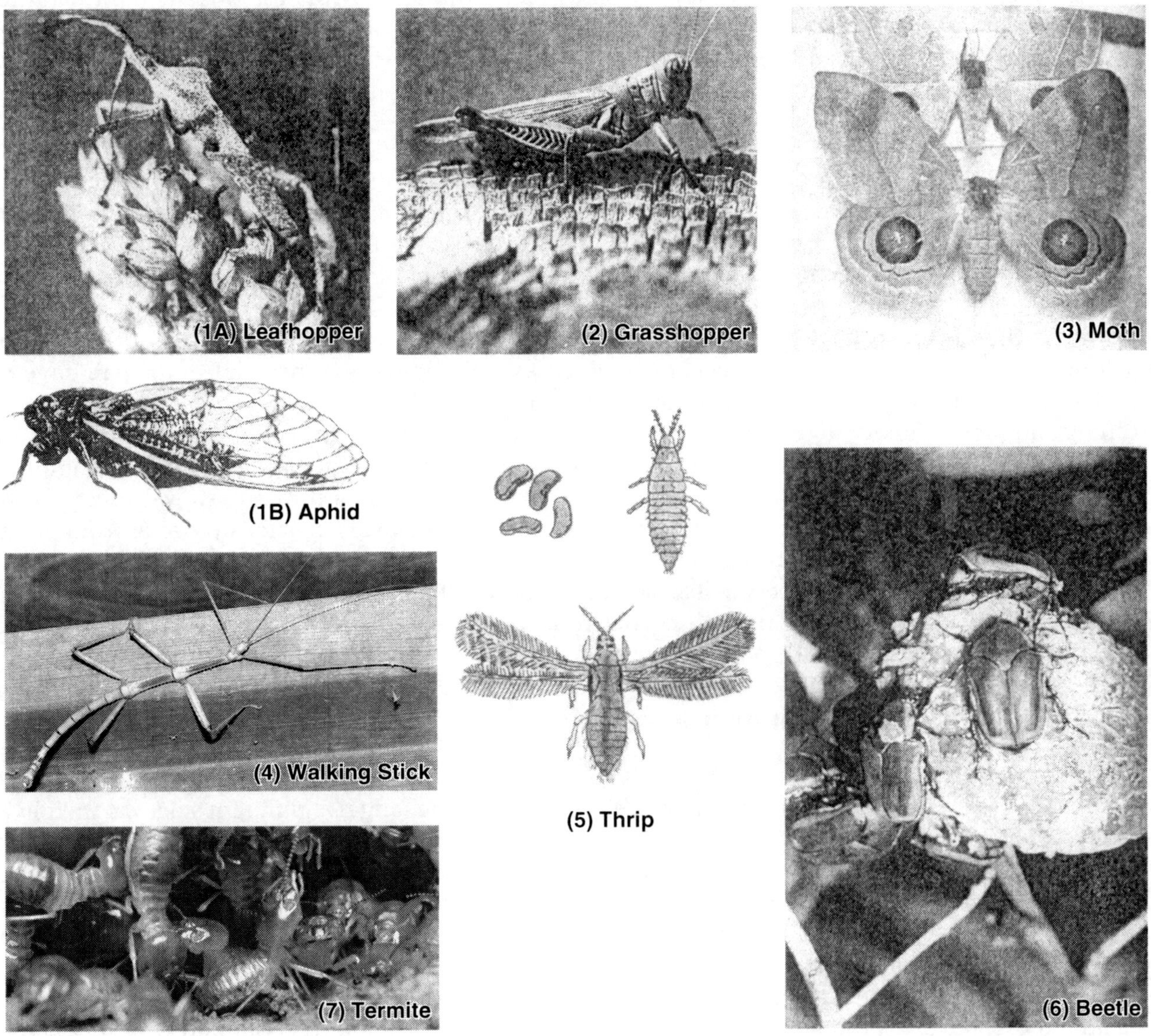

Figure 18–1

The orders of insects that damage living plants or forestry products include:

1. ***Hemiptera*** include the true "bugs," such as leafhoppers and plant bugs. Sternorrhyncha, a suborder of the *Hemiptera* (the former order *Homoptera*), contains whiteflies, scale insects, and the aphids.
2. ***Orthoptera*** includes the grasshoppers and locusts.
3. ***Lepidoptera*** includes the moths and butterflies.
4. ***Phasmida*** are the insects commonly called "walking sticks."
5. ***Thysanoptera*** is the order thrips belong to.
6. ***Coleoptera*** is the largest of the insect orders, which includes the beetles.
7. ***Isoptera*** is the order of insects that includes termites. An infestation of these insects can be very detrimental to the forestry industry.

Trees are like most other organisms; they are susceptible to attack by disease, insects, and parasites. Whole forests can be wiped out by an insect or disease infestation. The identification of the disease or pest is the first step in management and control methods. In this exercise, you will learn to identify insect and disease pests and some of the ways in which the problem can be controlled.

Mouth Parts

Insects that feed on plants are grouped by how they feed. This classification is determined by how the mouth parts of the insect are structured and how food is taken into their mouths (see Figure 18–2). Chewing insects tear off bits of plants using mandibles (jaws), which are strong enough to bite off plant parts and grind the material up by chewing. Examples of these insects are most of the beetles and grasshoppers.

Another group of insects have structures called **chewing and lapping mouths**, which are used to lap up liquid. Examples of this type of insect are bumblebees and honeybees. They have long, hair-covered, tongue-like projections that reach into blossoms and lap up nectar.

Rasping and sucking insects have mouth parts that scrape off cells and particles from a plant surface and suck the torn-off pieces into their mouth. Thrips feed on plants in this manner and belong to this group.

Some insects have mouth parts that are shaped like and function like a drinking straw. Insects in this group, called siphoning insects, extend this mouth part, called a **proboscis**, to suck up nectar from plants. The moths and butterflies belong to this group.

Some of the most damaging insects make use of a proboscis that pierces a hole in the outer layer of a plant surface and sucks the sap from the plant. The group with these mouth parts is called **piercing and sucking**. Scale insects and aphids belong to this group.

Flies use a mouth structure that serves as a sponge to absorb liquid nourishment. Nonbiting flies belong to this group, which is labeled **sponging insects**.

Life Stages

Insects differ from other living organisms in that their physical form changes at different stages of their lives. These different stages are considered phases in the insects' entire life cycle. Insects go through changes as they move from one stage into another. This process is called **metamorphosis**. Not all insects go through a metamorphosis; some, like the silverfish, merely go through a process of molting in which the insect outgrows the outer covering, sheds it, and then grows a new one. Most insects, however, go through some form of metamorphosis.

In a complete metamorphosis, insects go through four stages from the time they hatch until they are mature adults capable of reproducing. At each of these stages, the insect looks completely different than it does at the other three stages. When the young insect hatches, it is usually in the larval stage, which means that it has a soft tubular body and looks very much like a worm. The larva is usually a voracious

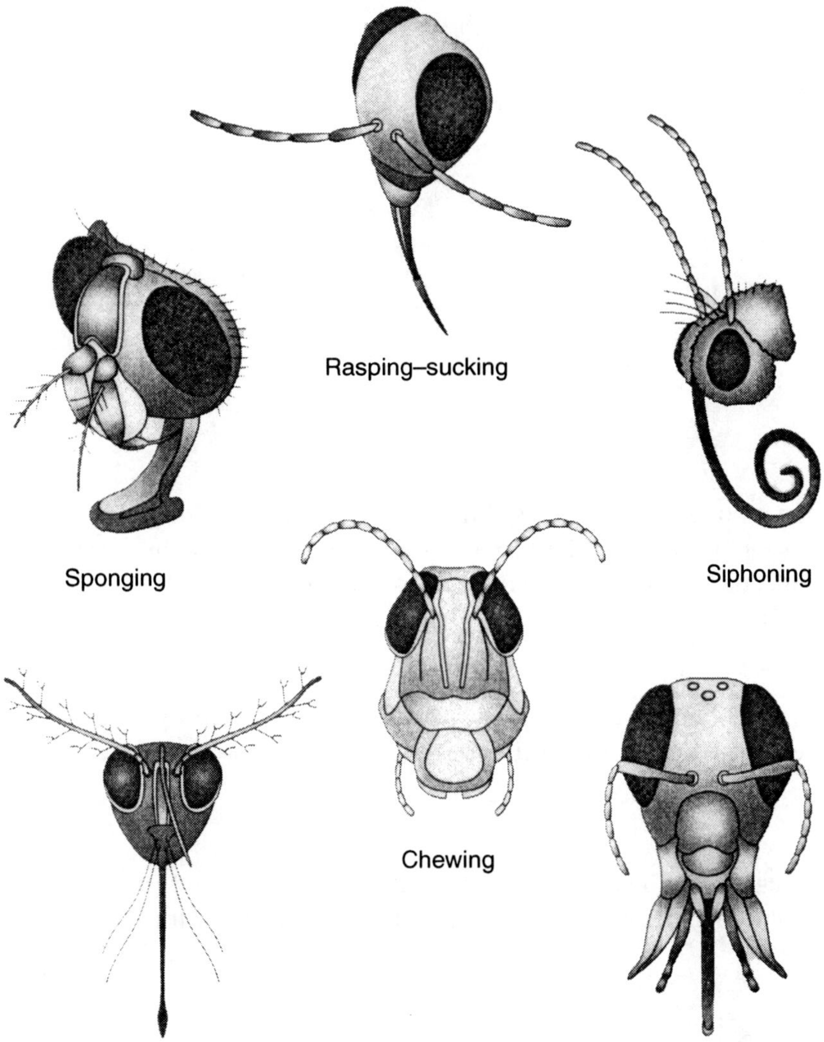

Figure 18–2

eater that can do a lot of damage to plants. When the larva matures, it passes into the pupal stage, which is usually a relatively dormant stage. During the pupal stage, the intermediate stage between the larva and the adult, the body tissues of the young insect is converting from a larva into an adult, which is its final stage. At this point in its life cycle, the insect reproduces and lays eggs. When the eggs hatch, the cycle begins again.

Some insects that spend at least part of their lives in water undergo an incomplete metamorphosis. In this process, the insects go from an egg to a form called a **naiad**, which differs completely from the adult in looks and function. When the growth process of the naiad is complete, the naiad changes to a winged adult. Examples of this type of insect are damselflies and dragonflies.

Insects that undergo a gradual metamorphosis hatch from the egg into a **nymph**, which looks very much like the adult. The nymph has different feeding habits and may live in a different place than the adult. From the nymph stage, the insect goes to the adult stage. Grasshoppers and cockroaches are examples of insects that undergo gradual metamorphosis.

Materials Needed

- Small bottles
- Insect identification guide
- Notebook and pencil
- Dissecting microscope
- Straight pins
- Styrofoam squares (2 feet × 1 foot)
- Paper labels

Procedures

Step 1

On a warm day, select a wooded area and carefully search for insects. Turn over fallen branches, logs, rocks, and other debris. Also look on trees, under leaves, in flowers, and on bushes, grass, and weeds. Be careful in handling the insects because many can bite or sting. Wearing gloves will help. Avoid touching the insects directly whenever possible.

Step 2

Back in the classroom, anesthetize your insects. This can be done by simply placing them in the freezer for a few hours. Using the straight pins, attach the insects to the foam board. Space them evenly, leaving room for labels.

Step 3

Using an insect identification guide or the Internet, identify each insect, its current life stage, mouth parts, and whether or not it is beneficial or harmful to plants. Record this information in the chart provided on page 145.

Step 4

Now that you have identified each insect, add a label to your collection. A proper label should include the scientific and common name, location collected, date collected, the collector's name, and any other important information. In this lab, the important information will be, whether the insect is "harmful or beneficial to plants." The label should be placed under the appropriate specimen.

Hypothesis

After reviewing the procedures above, predict in the form of a hypothesis what kinds of insects you think you will find.

____

____

____

Observations

Using the following chart, analyze the insects you have collected.

Insect	Life Stage	Order or Type	Mouth Type	Type of Damage	Beneficial or Harmful

Conclusions

Refer back to your hypothesis. Did the specimen you collected support your hypothesis, or will you reject it? Explain.

____

____

____

Chapter 19

ANATOMY AND CLASSIFICATION OF TREES— *TREES HAVE CLASS*

OBJECTIVES

After completing these lab exercises, students should be able to:

- discuss the various means of identifying trees
- identify ways of classifying trees
- list the parts of a plant cell
- identify the structures within a cell
- explain the function of cell organelles within a plant cell
- identify the types of leaf tissue cells
- explain the functions of the different types of leaf tissue cells

EXERCISE 1 SPOT THE DIFFERENCE—DISTINGUISHING CHARACTERISTICS

Background

Most plants, including trees, have physical characteristics that make them somewhat similar to other plants. In spite of their similarities, however, each of the species of plants and trees is distinctly different from all of the others. Examples of these differences that are easily observed include size, shape, and color. There are also some very subtle differences between closely related plants that can be distinguished only through careful observation.

Materials Needed

- Leaves from 20 trees
- Textbook, *Introduction to Forestry Science*
- Tree identification guide

Note: Find samples that have part of the tree's branch, a seed, and a cone, if applicable. You may also want to use these samples for Chapter 20's lab exercises.

Procedures

Step 1

From your area, collect your leaf samples from each of 20 different trees.

Step 2

Using Figure 19–1 through Figure 19–9, determine what kind of leaf structure each leaf has and list it in the chart provided. The leaf types are: needle-leaf conifer, scale leaf conifer, untoothed simple, toothed simple, lobed simple, and compound leaves. Compound leaves come in one of three arrangements, which are: pinnately compound, bipinnately, and palmately. Be sure to distinguish what kind of compound leaf your samples are.

Needle-Leaf Conifers

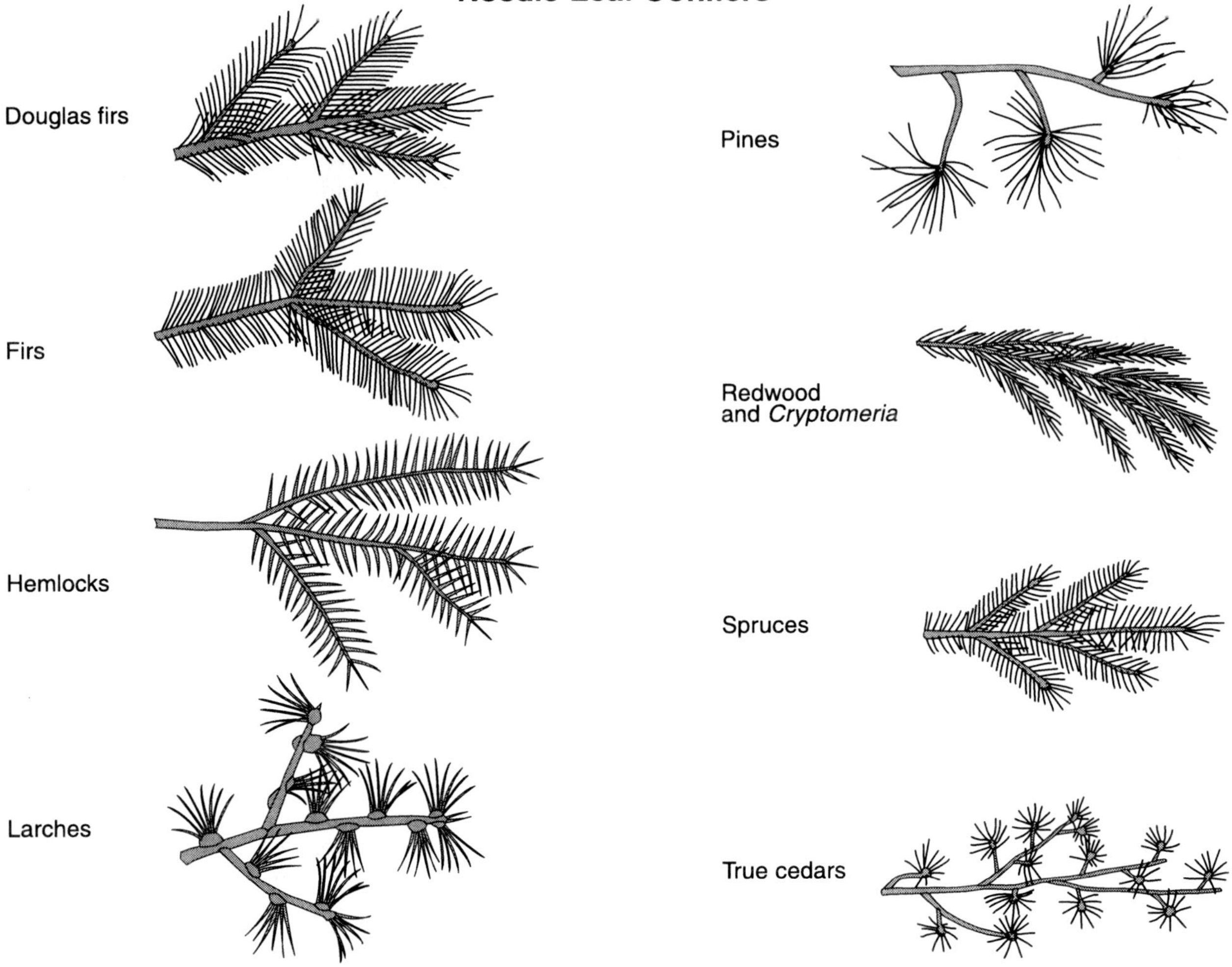

Figure 19–1

Needle-Leaf Shapes

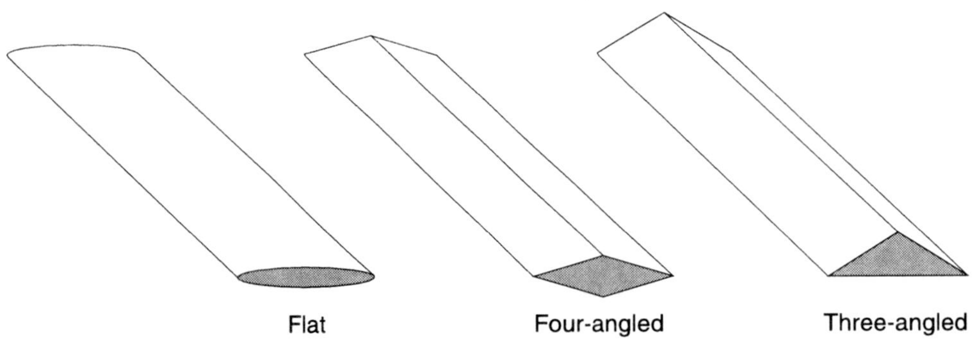

Figure 19–2

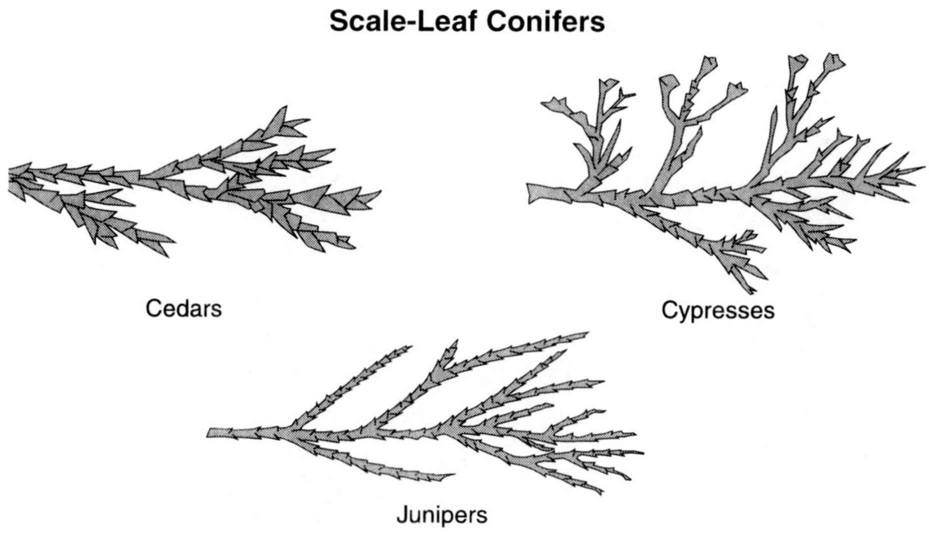

Figure 19–3

Figure 19–4

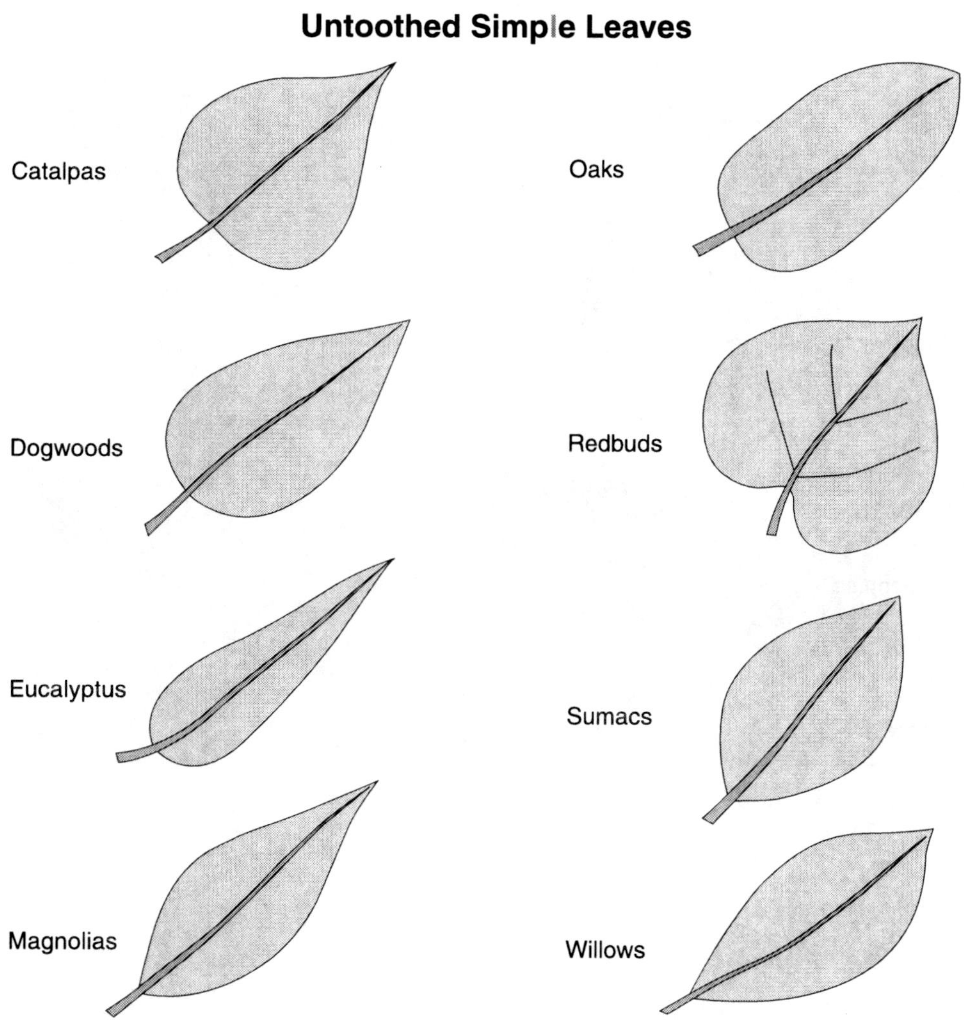

Figure 19–5

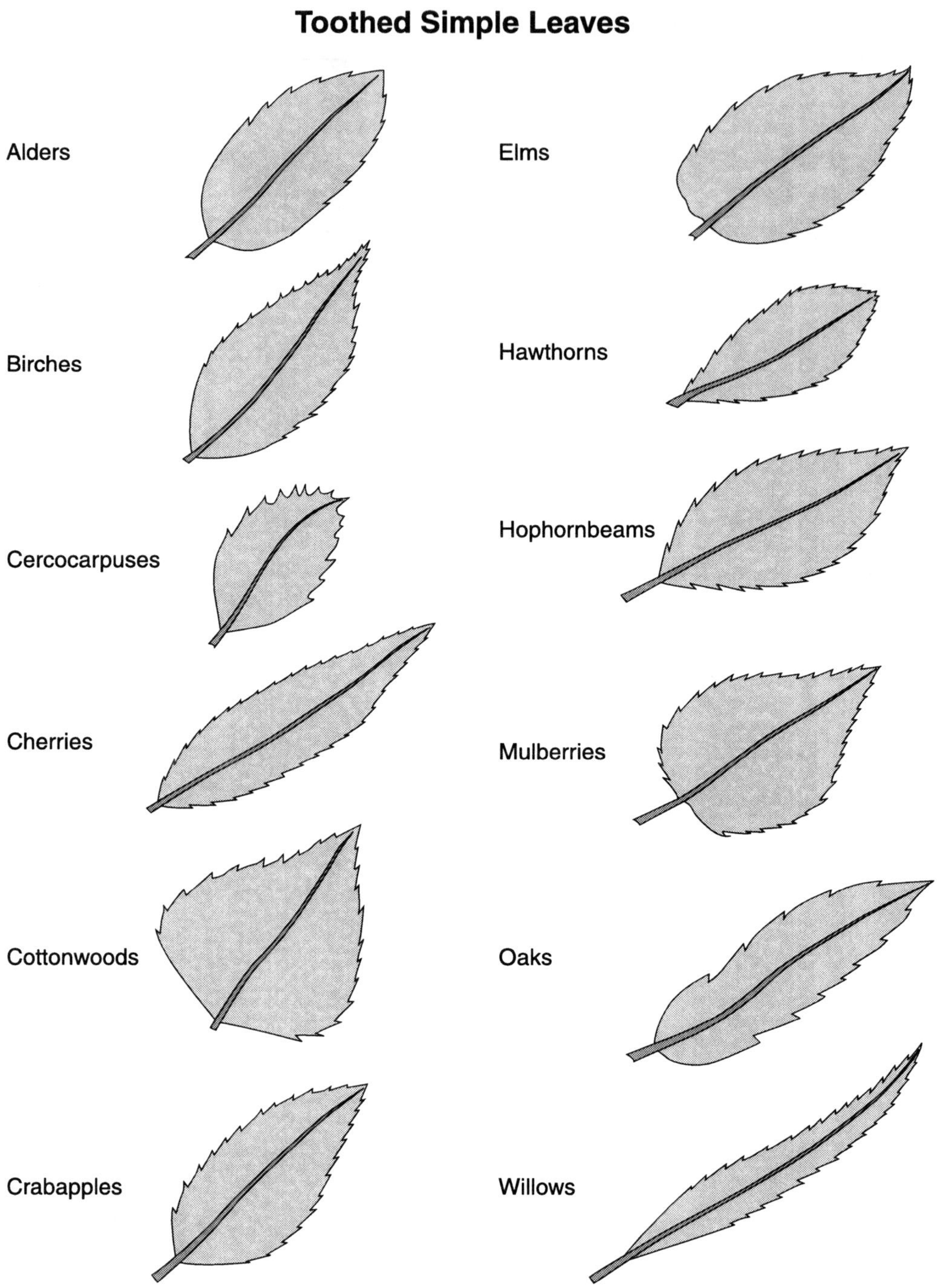
Toothed Simple Leaves
Alders
Elms
Birches
Hawthorns
Cercocarpuses
Hophornbeams
Cherries
Mulberries
Cottonwoods
Oaks
Crabapples
Willows

Figure 19–6

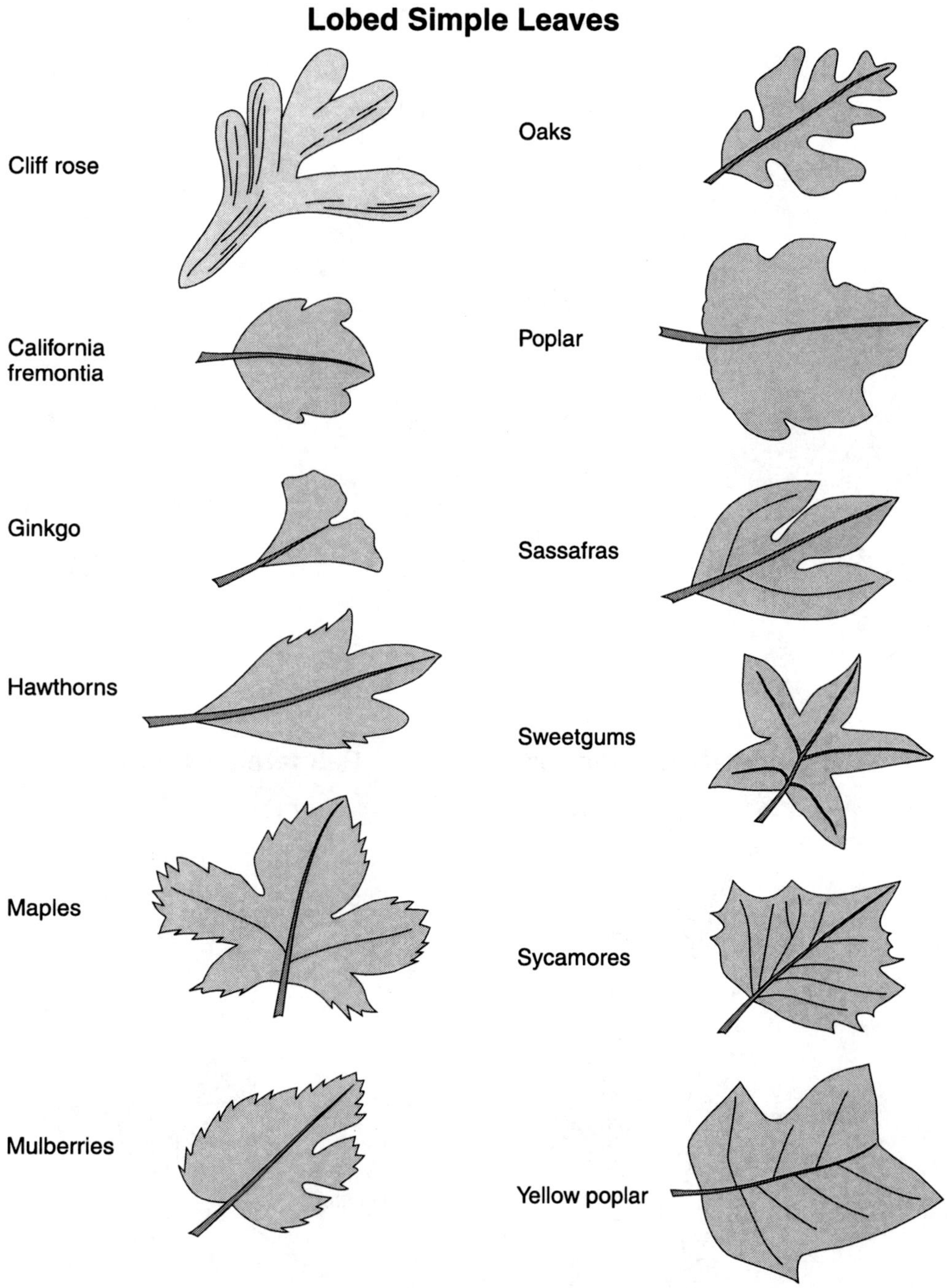
Lobed Simple Leaves
Cliff rose
Oaks
California fremontia
Poplar
Ginkgo
Sassafras
Hawthorns
Sweetgums
Maples
Sycamores
Mulberries
Yellow poplar

Figure 19–7

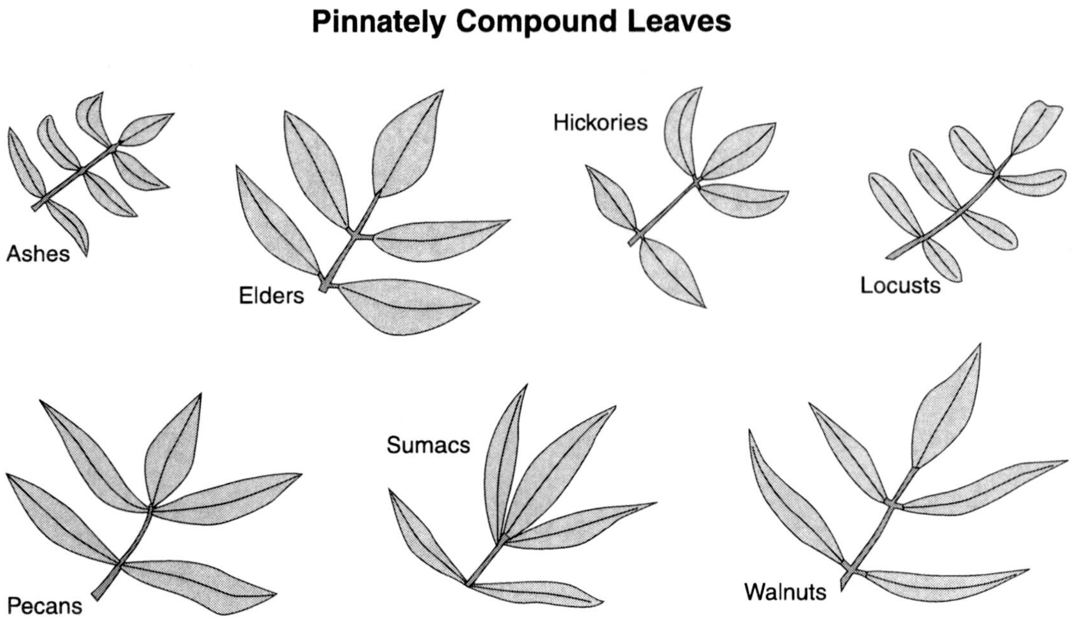

Figure 19–8

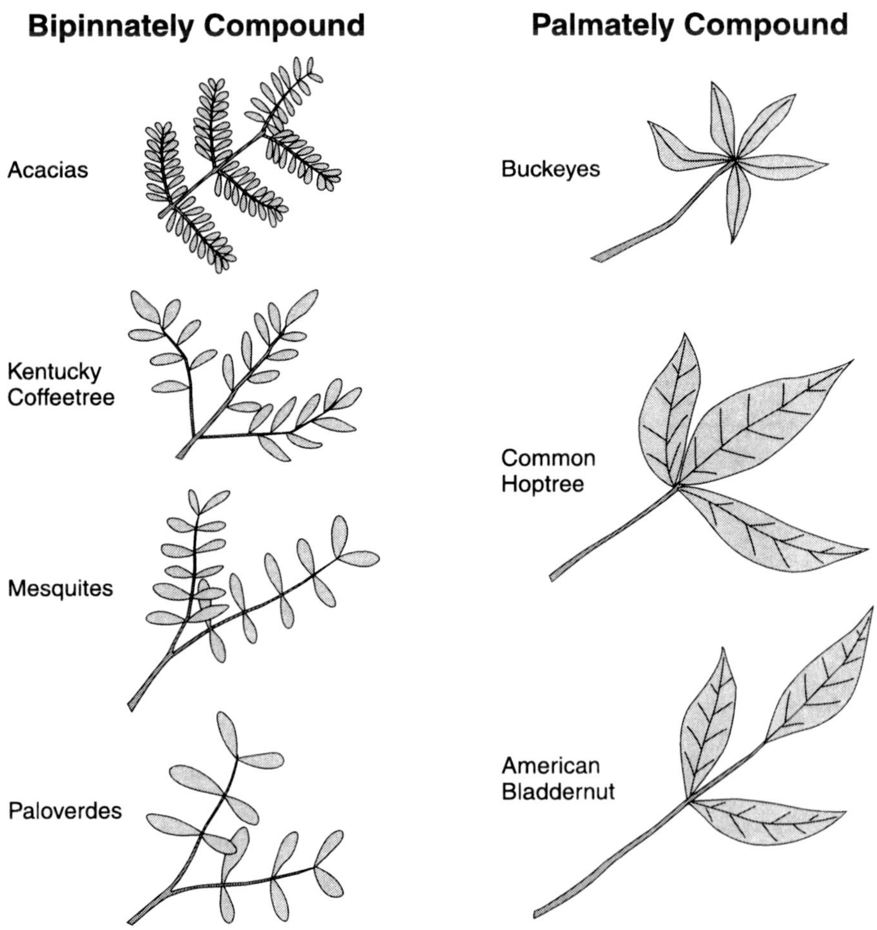

Figure 19–9

Observations

Leaf Sample	Type of Leaf
1	
2	
3	
4	
5	
6	
7	
8	
9	
10	
11	
12	
13	
14	
15	
16	
17	
18	
19	
20	

EXERCISE 2 I'M FULLY EQUIPPED—LABORATORY EQUIPMENT MISSION

Background

In the study of the sciences, you will often find yourself learning or working in a lab setting. It is important for you to become familiar with some of the equipment typically found in working laboratories. In this exercise, you will spend a little time using common tools that **botanists** (plant scientists) use every day.

Materials Needed

- Microscope
- Beaker
- Slide and cover slip
- Eye dropper
- Test tube

- Graduated cylinder
- Dissecting microscope
- Methylene blue
- Metric scale
- Biltmore stick
- Increment borer
- Pliable measuring tape
- Gold Star stickers

Procedures

Step 1

The classroom will be set up with common lab equipment from the list above. Your mission is to visit each item and complete a challenge in the time given and to answer the questions below. If you succeed, you will earn a *gold star* and move on the next challenge. If you fail, you will still have to move on, but will have to live with the knowledge of not fulfilling your mission. This message will not self-destruct. Good luck!

1. ________________ is used to dye an object to make viewing it better under a microscope.
2. Use this item to estimate the height of a tree. ________________________
3. These items are used by placing a specimen in-between them like a sandwich: ________________________.
4. When a small amount of liquid is needed, use a/an________________.
5. This item is used for many purposes. It can hold liquids, or it can be used in an incubator to culture samples. One thing it cannot do is stand on its own. ________________________.
6. This equipment magnifies items that cannot be seen by the naked eye: __________________.
7. If you were asked to add 100 mL to a solution, you would use this: ____________________.
8. If you needed to add 27 g of sodium bicarbonate to acetic acid, you would use this piece of equipment: ______________________.
9. The tool you would use to measure the diameter of a tree is: ________________________________.
10. This item magnifies small object so they can be viewed better: ______________________________.
11. This item is used to hold or mix solutions in: ______________________________.
12. To find out how old a living tree is, you would use a ____________________________________.

EXERCISE 3 BACK TO THE BASICS—PLANT CELLS

Background

Cells are the building blocks for living tissues. Within each cell are structures called **organelles**, which contribute to the growth and carry out all life-sustaining functions of the cell. Without these structures, the cell could not live and, therefore, the plant could not live. A basic understanding of the cell, its structures' functions, is essential to understanding how plants live, grow, and reproduce. Cells have different functions within an organism, yet all plant cells have the same organelles. The reason cells have the same structures yet can perform different jobs is the holy grail of **microbiology**. It basically comes down to signals being sent from the DNA to individual organelles, which then respond in a specific way.

Materials Needed

- Microscope
- Slides and coversheets
- Eye Dropper
- Elodea plant (found in the aquarium section of a pet shop)

Procedures

Step 1

A diagram of a plant cell is provided in Figure 19–10. Identify the structures within the cell, called organelles, and explain their functions in the chart provided on page 156. You may want to refer to Chapter 19 in your text.

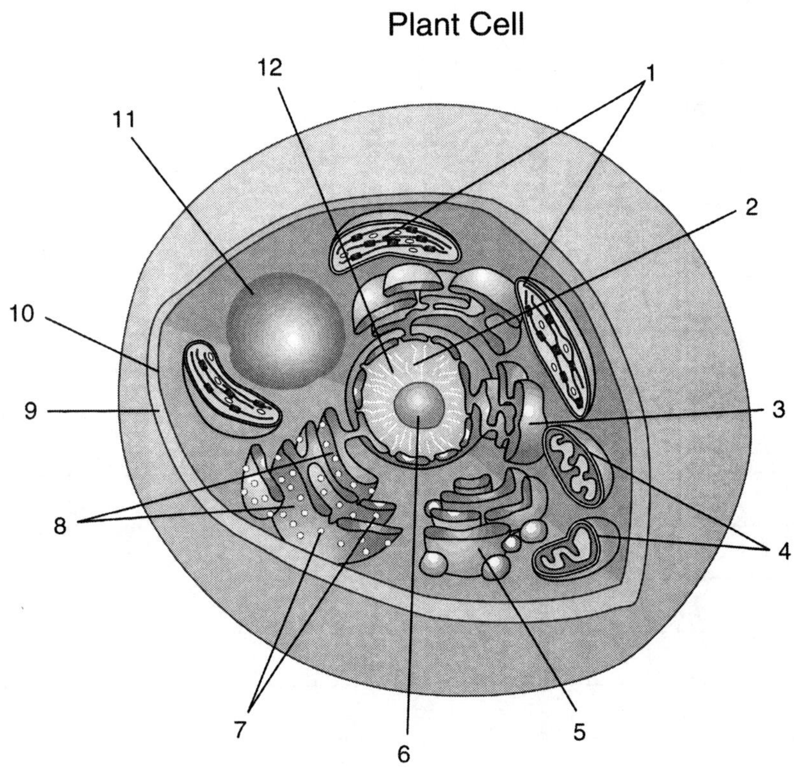

Figure 19–10

Step 2

A drawing of a cross section of a leaf depicting the different types of leaf tissue cells is provided in Figure 19–11. A **tissue** is a group of cells that perform the same function. In the chart provided on page 157, identify the name of the leaf cell tissue type and its functions.

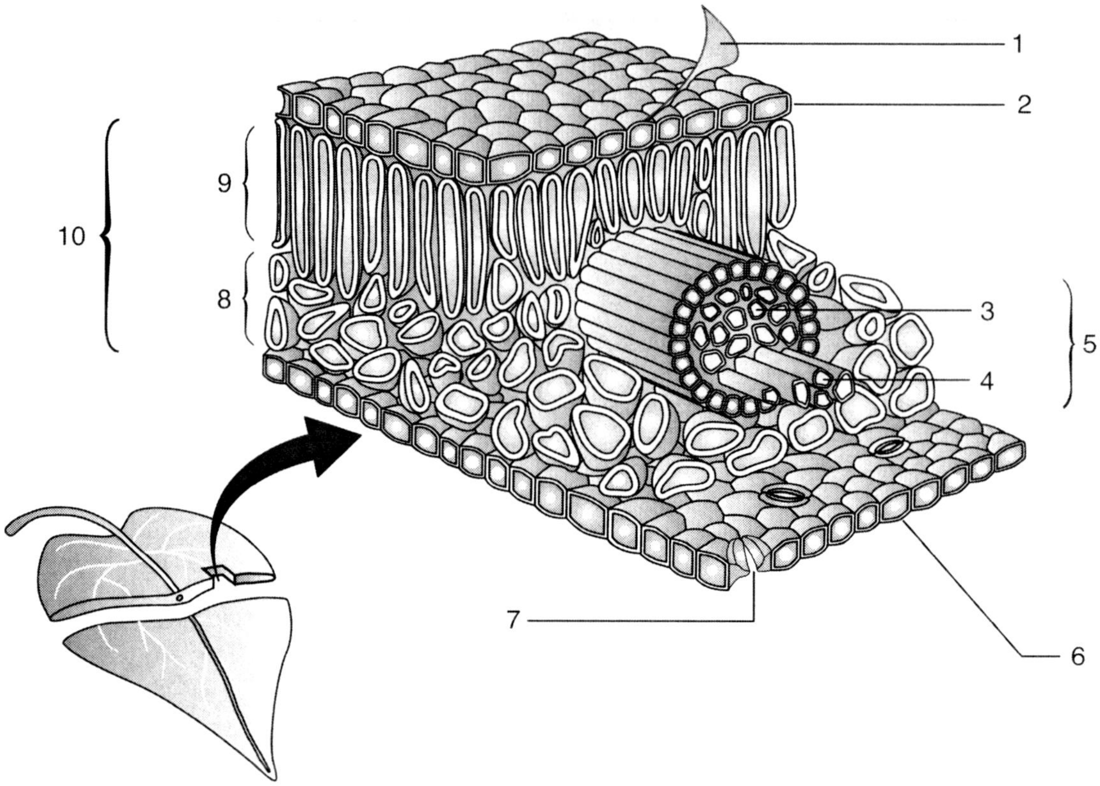

Figure 19–11

Observations

Exercise 3, Step 1—Plant Cell Structures

	Plant Cell Structure	Function
1.		
2.		
3.		
4.		
5.		
6.		
7.		
8.		
9.		
10.		
11.		
12.		

Exercise 3, Step 2—Leaf Cell Types

	Leaf Cell Type	Function
1.		
2.		
3.		
4.		
5.		
6.		
7.		
8.		
9.		
10.		

Step 3

Take a single leaf of the Elodea plant and place it on a slide. With the eye dropper, put a drop of fresh water on top of it. Gently place the slide cover over the leaf. Place the slide onto the microscope stage and bring the object into view. Increase the magnification until you can see the plants cells. In the space below, draw a picture of the cell and label the organelles and structures you see.

Elodea Plant Cell

Chapter 20

TREE IDENTIFICATION—*YOU LOOK FAMILIAR*

OBJECTIVES

After completing these lab exercises, students should be able to:

- use an identification key
- identify 20 trees from the student's home area

EXERCISE 1 YOU LOOK FAMILIAR—IDENTIFYING TREES

Background

Being able to identify and classify trees is an essential part of the study of forestry. It is difficult to communicate with other people about trees if you cannot call them by their common and/or scientific names. Trees are identified using various methods such as by leaf shape and type, bark texture, twig shape, tree size, hardwood or softwood, evergreen or deciduous, and by scientific name.

Materials Needed

- Samples collected for use in Chapter 19 or 20 new leaf samples

Note: Find samples that have part of the tree's branch, a seed, and a cone, if applicable. You may use the samples you collected for Chapter 19's lab exercises.

Procedures

Step 1

Collect your samples from 20 different trees in your area. Be sure to collect tree samples from both deciduous and coniferous trees.

Step 2

Using the Tree Identification Chart provided for this exercise, specify the identifying characteristics of each leaf and tree. After identifying the characteristics of the tree, identify its common and scientific names.

Observations

Tree Identification Chart

Leaf Shape	Bark Texture	Deciduous or Evergreen	Softwood or Hardwood	Common Name	Scientific Name
Lobed simple	Rough	Deciduous	Hardwood	White Oak	*Quercus alba*

Leaf Shape	Bark Texture	Deciduous or Evergreen	Softwood or Hardwood	Common Name	Scientific Name

EXERCISE 2 KEEP IT FOREVER AND EVER—PRESSING AND MOUNTING TREE SAMPLES

Background

Preserved samples of tree leaves, seeds, and branches allow foresters and plant scientists to learn and remember plant characteristics. They also provide a reference that can be called on in the future when questions arise.

Materials Needed

- Samples collected for use in Chapter 19 or 20 new leaf samples
- Newspaper
- Heavy books or peg board cut the size of the paper
- Large heavy paper
- Glue
- Pen
- Laminator or clear shelf paper

Procedures

Step 1

As you collect the leaves to be used in Exercise 1, collect a piece of the branch and a seed or cone from the tree. Press the leaves by placing them between layers of newspaper and piling heavy books on them. You may use a plant press if available. After several days, the leaves should be removed, placed on poster board, and glued down. After the glue has dried, the boards should be laminated or covered in clear adhesive shelf paper, which is available at most hardware and home stores. If the seed and branch samples are small enough or flat enough, they may be laminated or preserved under shelf paper as well. If not, you may attach them after covering the leaves.

Step 2

Create a label for each specimen, listing the:

- Common name
- Scientific name
- Bud arrangement (determine whether this is, i.e., alternate, opposite, or whorled)
- Names of any seed structures you can identify, such as the samara of maples and the catkins of birches that you may have collected

Step 3

Prepare the samples neatly and accurately. It may be possible to enter these displays in local fairs and youth shows.

CPSIA information can be obtained
at www.ICGtesting.com
Printed in the USA
FFOW02n0642190814
6942FF

9 781111 308414